W9-COH-343

*S*tories and Essays for

Sleepless Nights

Edited by
Elyse Cheney and Wendy Hubbert

THE *LITERARY*

INSOMNIAC

(Doubleday

New York   London   Toronto   Sydney   Auckland

PUBLISHED BY DOUBLEDAY
a division of Bantam Doubleday Dell Publishing Group, Inc.
1540 Broadway, New York, New York 10036

DOUBLEDAY and the portrayal of an anchor with a dolphin are
trademarks of Doubleday, a division of Bantam Doubleday Dell
Publishing Group, Inc.

Book design by Maria Carella

Library of Congress Cataloging-in-Publication Data
The literary insomniac : stories and essays for sleepless nights /
    edited by Elyse Cheney and Wendy Hubbert. — 1st ed.
        p.   cm.
    1. Insomnia—Literary collections.   2. Night—Literary
collections.   3. Sleep—Literary collections.   4. Prose
literature—20th century.   I. Cheney, Elyse.   II. Hubbert, Wendy.
PN6071.I58L58   1996
808.8'0353—dc20   96-23022
                                                              CIP

ISBN 0-385-47771-6

Printed in the United States of America

December 1996

First Edition

10   9   8   7   6   5   4   3   2   1

Contents

INTRODUCTION

The world is divided into two groups, claims Iris Murdoch: those who sleep, and those who do not. Indeed, insomnia is so widespread among writers that it seems almost a prerequisite for literary success. Vladimir Nabokov had such trouble sleeping that he kept a separate bedroom so as not to disturb his wife. The Brontë sisters spent the wee hours sleepwalking around their dining room table. Mark Twain went so far as to warn his friends against going to bed—"so many people die there."

What keeps millions of Americans awake at night?

No one is better suited to answer this question than writers. And in these pages, the literary voices of our time explore the fears and passions evoked by the night. You'll hear a wealth of perspectives, including Mark Richard's dark rant against the noises of the darkness; Annie Proulx's experience of waking up in the middle of the night; Siri Hustvedt's eerie story about a young Barnard student who is hired by a doctor to ferret through his dead wife's things. You'll find great advice on coping with insomnia from Quentin Crisp; Tim Cahill's hilarious account of his nocturnal escapades in Spain, and George Dawes Green's philosophy on "The Bondage of 24."

Sleep is rife with symbolic associations, bringing up issues of identity and death. Sleep punctuates the hours, arousing the demons that we shut out of the daytime through work and play and living. During the night, however, our subconscious brings to the surface fears and fantasies often too dark or frightening for us to comfort ourselves.

Whether we suffer from chronic insomnia or from the occasional troubled night, restlessness and uncertainty describe our times. Cultural critic Harold Bloom has called this decade the "age of anxiety," and *The Literary Insomniac* comes out of this societal temper. It is a catalog of experiences and a way of understanding a massive cultural phenomenon. While the stories and narratives in this volume always entertain, they also reflect the vicissitudes of the human condition.

Insomniacs suffer not only during the night but through the following day as well, unable to concentrate or to perform energetically. Even an hour of waiting to fall asleep can cause extreme anxiety, as the day's worries race through the mind with irritating repetitiveness. Like sufferers of any condition from lovesickness to loneliness, insomniacs want to know that they are not alone in their experiences.

We hope you'll find comfort and understanding within these pages.

Elyse Cheney and Wendy Hubbert
New York City, 1996

Thomas Beller

Music at Night

1.   I first saw him at the Little Apple grocery, just across the street from my building on 93rd Street and Amsterdam Avenue. It was a winter morning. I had just woken up with a hangover and was negotiating with some difficulty the narrow cluttered aisles of the store in search of an orange juice. The Little Apple played salsa music at such a volume that entering the store was like ducking into a tiny Latin nightclub with a canned-food theme. Unfortunately for me, I often visited shortly after waking, and so experienced the store as a huge walk-through alarm clock. My state of mind on this occasion was jolted, bleary, and a bit dim.

He was buying a beer. He was at the cashier, and I thought he was paying; after a moment it became clear that he was just standing there chatting with the cashier in Spanish. I took in the salient details one at a time: Man with sixteen-ounce Budweiser in brown bag, opened (small head of white foam rising up, moistening edge of brown bag). Speaking Spanish. A bit on the smallish side, with a paunch. Leather jacket. More Spanish. A ponytail. More Spanish. A wispy goatee. Black leather boots.

I finally stepped forward, realizing this was not a line, and paid for the orange juice. I received a sympathetic glance from the cashier, a burly man with a mustache who is just the sort of person you want to buy orange juice from at eleven in the morning when you are extremely hung over, because he looks as though he may be hung over as well, though with a shower. Then, in turning, I grasped that this was a particular kind of black leather, which is to say, it had its own stylistic message. It was not Fashion Black Leather or Punk Rock Black Leather or S&M Black Leather. It was *Motorcycle* Black Leather. I vaguely took this in and then picked up some strange information about his left arm. Just a glimpse. Something seemed wrong with it; it hung stiffly, bent at the elbow against his chest. And then out the door and into the relative salsa-less calm of Amsterdam Avenue.

And that's how he came to be registered—a Hispanic man who wore a black leather jacket and faded blue jeans and had a goatee and long hair and a withered left arm. Not a known quantity, but a vaguely recognizable one; one of those images that go into that vast bank of knowledge one instinctively keeps in the hopes of making the world a more familiar and slightly manageable place.

• • •

2.   There was a lot of sound in my neighborhood. There was a bus that ran up Amsterdam Avenue, and quite a few trucks bumping along, and now and then a car would honk furiously at some pedestrian laconically making their way across the street with a hit-me-if-you-dare look on their face (there was a lot of this, and once someone did in fact get hit). In the summer there was Mr. Softee's jingle chiming sweetly in the distance, the aural equivalent of rainbow sprinkles.

But mostly what one heard was music. Brief snatches of music. Music from radios. Music from car radios. Music from fast-moving car radios, which provided the additional complication of the Doppler effect—music heard coming and going. And mostly, the music was at night.

It's a strange jumbled neighborhood, a poor cousin of the Upper West Side that has become buffed and polished over the last decade and a half. That neighborhood ends precisely at 92nd Street (as of this mid-1990s writing), where Broadway and Amsterdam Avenue, both of which have been on a continuous upward ascent starting at around 72nd Street, take a precipitously downward dip. If you walk across Amsterdam and 92nd Street and stop in the middle of the avenue, you get a kind of urban vertigo—the avenue stretches away in a long straight line to the south, disappearing at the lip of midtown, somewhere in the Sixties, and to the north it begins a long downward swoop in preparation for the frantic undulation of huge hills that marks the beginning of Harlem. Culturally and economically speaking, 92nd has no special claim, but the topography makes it a dividing line of some kind.

I lived on 93rd Street and Amsterdam, on the east side of the avenue. A good example of the uptown half of the Upper West Side would be 93rd between Amsterdam and Broadway: It slopes downhill toward Broadway at a sufficient degree so that going down one is loose-limbed and bouncy and going up is a distinct trudge.

If you walk on the south side of the street you will pass a white-brick apartment building with no doorman, about six stories high. This building housed three stores—a dilapidated television and radio repair shop which could easily have been confused with an indoor junkyard, except that there were invariably eight or ten televisions turned on at once, and a tiny candy store called Nuts and Dried Fruits, and an even tinier dry-cleaning establishment run by an old man who never seemed to have any customers and spent his days reading Chinese newspapers and smoking cigarettes (I knew this because I was one of his few customers). Other than these three stores, the building's main distinction was that a huge chunk of rock once fell, unprovoked, from its upper reaches and sat for two days on the sidewalk, like a corpse. It looked like a crime scene, and when I first encountered it, several minutes of staring at the chunk of rock and the police barrier surrounding it elapsed before I grasped what had happened—an understandable lapse, I think, because one of the central pillars of sanity in New York is the firm belief that large chunks of building will not spontaneously detach themselves and crash to the sidewalk below.

Past this humble building was a hulking structure that, upon examination, revealed itself to be a synagogue. I had never seen a single person enter or leave it, but sometimes the metal gate across the front was ajar, and the little signs announcing this or that sermon changed with regularity, so someone was making use of it. There are many old ornate synagogues secreted around New York that, if stared at for a minute, reveal themselves to be beautiful structures, but this was not one of them.

Finally, at the corner of Broadway, there was an enormous

prewar doorman building. On the opposite corner of 93rd and Broadway was a church. The religious activities of this church had never been apparent, but there were AA meetings in the basement, and in the summer they opened the windows and one could hear them clapping. Looking down through the open windows one saw a lot of laps and feet and hands—the hands held smoldering cigarettes or little plastic cups of Kool-Aid, or just clasped each other. I have vivid memories of all these disembodied hands and laps, fluorescent-lit, that belonged to former alcoholics. Other than this church, the rest of the south side of the block was taken up by a housing project.

Walking down from Amsterdam to Broadway on the north side of the street was an entirely different experience than the south side (and I should add that after that huge chunk of concrete and brick landed on the sidewalk, walking on the north side was a damn regular experience). There are many different kinds of public housing and, to judge from some of the other buildings scattered around the neighborhood, some of them are quite pleasant, even a bit luxurious.

This was not an example of luxury public housing. Not that it was squalid. It was not. But it was very big, very ugly, and, it appeared, a bit overcrowded. The people who lived there were not destitute, not wretched or hungry. But not for a moment to be confused with the well-groomed types whose heels (men's and women's) clacked down the sidewalk every morning en route to the subway from the brownstones or co-ops that populated 93rd Street from Amsterdam to Central Park West.

The project was a huge brick structure that someone, in a very misguided attempt at beautification, had decorated with turquoise squares running up the side of its sixteen stories, like Morse code. In the right context, every color can be beautiful. This was not the right context for turquoise.

The building itself was set back from the sidewalk by about thirty or so feet, in which there was a tiny bit of public space, some concrete chairs and tables, a cross between a playground and a tiny park. The main effect of this space was to put distance between the sidewalk and the building's front door. The front door was a slab of dented metal that seemed preposterously small in relation to the building, like the hatch of a tank or the entrance to a beehive. And there was always a gang of people hanging around it. Sometimes it was a gang of two, sometimes twenty. They usually looked out at the people passing by on the sidewalk with the serene indifference of people staring at a slow-moving river. At two or three in the morning the cluster out front seemed particularly conspicuous, which is to say, the river of pedestrian foot traffic has been reduced to a single distinct droplet—you—and their presence was cause for a certain amount of self-consciousness.

They usually had a radio. Sometimes it was loud. Sometimes soft. But there was always some sound coming from that space around the project's battered metal door.

3. The Heavy Metal Puerto Rican—which is how I came to think of him, though he could have been from Nebraska or the Dominican Republic or Spain, for all I knew—became a familiar fixture in front of the project and around 93rd Street. I had been living there for a couple of years by the time I first noticed him, and was never completely sure if he had been there all along and I simply noticed him for the first time that morning in the Little

Apple grocery, or if he had recently arrived. I suspected the latter; he was too incongruous in the neighborhood to go unnoticed. His clothes and, to an extent, his demeanor suggested knowledge of other worlds that had nothing to do with 93rd Street and housing projects; they suggested wide-open spaces, long straight lines of pavement that tapered to the horizon, gleaming chrome and chopper motorcycles. And yet there he was hanging out, beer in hand (right hand), with everyone else in front of the dented metal door. He hung around with the ease of someone who was on their home turf. Vague questions began to form in my head. His age was unclear, but he was no younger than thirty and could well have been in his forties. Maybe he saw *Easy Rider* when it came out and it changed his life. Maybe he had been on the road for decades, roaming over the interstates with his gang, or maybe just alone. But surely he had been a biker, a wild guy who drank and caused trouble and had a motorcycle that woke people up when he gunned it at night. And then the accident, and the frowning doctors, and the cast. Maybe it went over his shoulder, his chest, maybe he was in bed for weeks, months, felt his muscles go slack, disappear, and then even when he could walk out of the hospital, his arm was still in a cast, and the dour faces of the doctors bearing their cold hard facts, and the odds of this, the chances of that, the muscles that would never heal, and finally the scrawny chicken bone of a limb emerging from the dirty plaster, signed and drawn on until there was no white left. And eventually that came off, to be replaced by the sling, some rope and string to keep the thing aloft and safe against his chest. And then back to projects on Amsterdam Avenue. Home? I couldn't begin to guess.

I began to unconsciously check the arm. I was checking for improvement. I somehow formulated a theory that when it got better he would move on, hit the road, join up with his old gang. If he seemed strange standing around in front of the project and watching people walk by, then he in all likelihood also seemed strange standing around with a bunch of paunchy white guys and their motorcycles listening to AC/DC, or Steppenwolf, or the Grateful Dead, or whatever bikers listen to.

Time went by, seasons changed, and the arm did not get better. I recall the summer, maybe two years after that first sighting in the Little Apple, that he emerged from the thousands of vaguely registered images that populated the neighborhood and became a distinct source of interest for me. For one thing, the end of winter meant that one could spend more quality time hanging around outside (there was winter hanging around, but the hangers were usually enveloped in enormous amounts of down clothing and looked like a congregation of Michelin tire men). For another, on this particular spring, the Puerto Rican biker brought his own radio. This was something of a revolution.

As I said, the neighborhood was filled with sound, with music, and usually it was hip-hop, or sometimes Smooth R&B Jams ("ballads," I feel inclined to clarify, as I just imagined the furrowed brows of people staring at "Smooth R&B Jams" and trying to conjure a sound). Usually the music was heard in small segments, because either the music was moving (somebody walking along with his or her radio, or a car) or I was moving past the stationary radio on the way from my apartment to Broadway or back.

So it was all snippets. And since rap music so often uses samples of other, older songs, there would be snippets within snippets. On several occasions I would be lying in bed, and one of those car radios that are so big the woofers take up the entire back seat of a Jeep would come booming up Amsterdam at high speed, and within the course of five seconds I would hear a distant sound, then a thunderous sound which would be recognizable as some old song being sampled within another song, then the thunderous sound would undergo the weird Doppler effect and change tone, then it would be distant, then it would be gone. And I would be lying in bed blinking, utterly distracted from what my thoughts were seconds earlier.

Sometimes one would walk down the street and there would be more than one radio on and they would be playing different things and it would be dissonant. But it wasn't culturally dissonant, because it was two different rap songs, or something like that. Then one day I walked down the street—it was summer, a balmy late afternoon with the sky over the Hudson River already turning a little pink—and heard among the cacophony a peculiar sound. I looked toward it and saw him with his new radio. The Heavy Metal Puerto Rican was listening to Jimi Hendrix. This was quite shocking. It took the peculiarity of his appearance and, literally, amplified it. It was a sound track to his outfit. It gave him a voice, for me. I had seen him, and now I heard him. It wasn't a huge stereo, not one of those essentially home stereos that happen to take batteries and have a handle. But it was big enough so that the television repair shop and Nuts and Dried Fruits and also the project across the street were now the visual accompaniment to "Foxy Lady."

It was that summer that I saw him often or, rather, thought of him often when I saw him. The arm was not improving. The sling it was in seemed to have diminished in size and scope, but I glanced at the hand and wrist and forearm and they all continued to seem withered. I wondered if maybe he was born that way.

It was a very nocturnal summer for me. I am in general a night person, especially so in summer, but that summer was a special case.

The Little Apple closed at 11 P.M., which in my schedule was when I started to want a post-lunch snack. There were the Chinese restaurants on 98th and Broadway that delivered until 2 A.M., and I often made use of last call. Then there were two groceries down on 93rd and Broadway (early on I made my allegiance with one and stuck with it) and often I would go out at around two or three or four in the morning for a midnight snack. I would amble down 93rd Street on the project side (which happened to be my side, and the side of my grocery store, but mostly it was the side that a huge chunk of rock had yet to fall on; walks on the south side of the street were reserved for particularly nihilistic moods) and spend some time, often rather drunk time, standing around staring at the surreal abundance of lemons (incredibly yellow, and with bright lights trained on them as if they were movie stars) jammed into the Korean grocery, all the more surreal because just across 93rd Street was another, even larger grocery, equally abundant and overflowing, so much so that the flower displays outside each store seemed to bristle forward toward each other, their bright petals menacing one another from across the street like rival gang members.

And then the trudge back up 93rd Street, carrying a plastic bag with my midnight snack. Walking past the project, I usually glanced at who was hanging around out front. One has to be careful and somewhat calibrated in one's glances in such situations, as you don't want to appear confrontational and lock into a stare with someone, which usually is a way to start a fight, but at the same time I always felt that it was almost rude to scamper coweringly by, which is what most people did, as though just by virtue of hanging out with a radio late on a summer night in front of a project these people had incriminated themselves and defined themselves as *bad*. So I glanced over, held my gaze longer on some occasions than others, and checked out who was doing what. I often wanted very much to exchange a comradely nod with someone over there. I wanted to do this with my man, the Heavy Metal Puerto Rican, and in fact attempted a few very subtle acknowledgments, so subtle they were essentially slightly modulated blinks. I think he may have caught them, and blinked back.

All of this precise calibration must make me sound so neuroti-cally obsessive that it is almost a miracle I could walk down the street, but this precision is all in retrospect; at the time I was just dreaming through all this, as I was in a way dreaming and rather drifting through my life. It took place in a neighborhood in which I no longer live, and the internal landscape has changed as well, and so it seems very far away. As it was happening, the Heavy Metal Puerto Rican occupied my consciousness for the exact time he was right in my sights. It's only now, looking back on it, that he looms up, a peculiarity, a question mark, a gem and also something of consolation, though I don't know why. The urgency of my feelings about him now didn't exist then, and even if they did I'm not sure I would have had the gall to approach him, say hello, talk to him, and find out his story.

I never did. He exists in my memory only as a figure, a pres-ence, someone whom I can make some general inferences about but no more. Even on the most simple level, I can't answer the question of who, exactly, he is. The only thing approaching a personal ex-change between us took place very late one night during that ex-ceptionally dreamy summer of staring drunkenly at the lemons.

It was hot. I was making my way down 93rd toward my gro-cery and grocer, a nice Korean man with thinning hair, a color photo of his son playing soccer back in Korea in his wallet (he showed me once), whose fingers were so dried out that he had to mush them into a sliced lemon every time he reached into the register to remove a dollar bill. I went to the grocery as much for entertainment as for food.

And so on that night, as I was merrily bouncing my way down the hill sometime around three in the morning for the purpose of staring at the well-lit produce, I passed the crowd hanging around the front door. The congregation possessed an unusual density. A furious hip-hop beat emanated from a radio somewhere within the group. There was no doubt something menacing about them, but also something nice—down near Broadway a few crackheads would be scavenging, panic-stricken, while some Koreans sorted fruit—tossing away the bad strawberries, putting the so-so strawberries on the bottom of the plastic container, and putting the luscious red ripe ones on top. It would be weird and fragmentary down on Broadway; up here in front of the project one felt a community. Among the throng stood my man, utterly at ease. I averted my gaze as usual, sauntering by in my role as casual, odd, and possibly acceptable white person—after all, it was after three in the morning on a weeknight and I was out here too. But, to be honest, I wasn't all that casual, because I was in the habit of staring at my shadow, brightly outlined by the streetlamp above, to check if anyone was coming up behind me.

On my way back, not more than fifteen minutes later, I heard something odd as I approached the project. It drifted over in the wind and sounded familiar. I recognized it as Led Zeppelin. From thirty yards away I heard it clearly. It shrieked out against the buildings. And the buildings, and the streetlamps, and the hulking anonymous synagogue across the street, and the cars, both dilapidated and fancy, parked on the street with their complex hypersensitive alarm systems and their rudimentary chains wrapped around and around the steering wheel, all seemed to recoil with embarrassment, as though to say: Wrong Song!

A predawn sweet spot of cool had emerged in the atmosphere, a whiff of clarity that would last until the sun came up and everything became sweltering again. In the midst of this stillness blared Led Zeppelin from in front of the project, and as I approached I saw that the crowd had entirely dispersed. There remained only two figures, heads bobbing slightly to the music. One was my biker friend, and the other was a black man who was mostly bald and stood beside him. I got closer. The sound increased. The flimsy white plastic bag containing a coconut Frozfruit swung limply beside me, like a white flag of surrender.

"Man," I heard the black man say. "This brings back old times."

Old times! Led Zeppelin brings back old times? How totally strange, I thought. I made my way up the hill and debated ways in which I might show my appreciation, and at the same time considered the ways in which anything I did might be misconstrued. If I smiled and nodded, would it seem like a false connection? After all, how annoyingly predictable that after years of Run DMC and Public Enemy it should be Led Zeppelin that provoked a smile.

I loved the sound. It was so loud against the quiet stillness of the streets, so incredibly loud. I wondered if my man had simply flicked a switch and blasted everyone else off the front porch. I sensed eyes on my neck, my legs. I nodded my head in a barely perceptible bob to the crumbling beat. I was abreast of them, then a little past. The singer's wildly hysterical shriek ripped through the air. The Doppler effect had taken place and already there was a slight shift in its tonality, making the barely perceptible change from something approaching to the realm of something receding. I considered that this would be as fine a time as any to acknowledge my biker buddy, to express some kind of affinity, to show support for his crazy enterprise of being himself. But it was too complicated—I was suddenly gripped by the fear that a mere glance would be taken as a sign of disapproval—and at last all I could manage was an involuntary tilting of my head. I let it roll back on my shoulders and I breathed deep with my eyes closed.

Just then the bright streetlamp that arched high above that bit of sidewalk blinked off. They call these things cobras, because of their shape, but it's also appropriate in terms of ambiance—poised, aloof, glaring. I never quite understood why these keepers of nighttime city light blinked off for short periods of time, as though in need of a rest; it always seemed completely arbitrary and at the same time fraught with significance. And now the illuminator of that stretch of city block took an illumination break. I could feel the change through my closed eyelids. I took several vertiginous steps with my eyes closed, and opened them to a different night than the one I had last seen. Above me swayed a few branches thick with leaves, and beyond them the thick black atmosphere. The darkness made the music seem like a dream. It echoed off all the hard surfaces of the street, surrounding me. I rued my own small failure to express my appreciation, but was consoled by what I considered the triumph of the Heavy Metal Puerto Rican. He was holding on to something. He was living his life adamantly. There was something brave and fierce and obstinate about it. The music screamed like a fire behind me.

"Been a long time since I rock-and-rolled," it said.

Tim Cahill

 REAMS

I was wandering the streets of Lima, Peru, in an exceedingly bad mood, primarily because I hadn't slept well in several nights. It was early evening in the fashionable Miraflores district when two men in front of me, both wearing suits and ties, began fighting over some business or another. One was propelled backward, directly into my arms. Involuntarily, and with absolutely no desire to get involved on my part, I clasped him, so that his arms were pinned at his side.

His opponent drew back a fist, then hesitated, contemplating, I imagined, the general tastelessness of striking a defenseless man.

"Gentlemen," I said sharply, in Spanish, "calm yourselves."

"What?" This also in Spanish, and from both men at once. "What?"

My Spanish lacks perfection, and it was possible that I had somehow verbally soiled myself.

Now there was a small crowd gathering, and the two antagonists were confused to the degree that their anger was completely defused. I released my man, who turned and politely asked me what I could have possibly meant by "calm yourselves."

I had one of those little electronic pocket translators. Several people were involved in the process by this time, and it was proven that I had not spoken the verb *"calmar,"* but had, instead, used *"calzar."* In effect, I had told the two angry men to go "wear yourselves like shoes."

It was street theater, over in a minute, but the man I'd grabbed fastened on to me and, for the next half hour, in the grand Latin tradition, he became my best friend on earth. What was I looking for? Could he help? The man was done with work for the day and he was at my service.

What I was looking for was unattainable. I wanted to sleep. I wanted to get into bed, turn off the light, and go to sleep. I don't sleep, I can't sleep, I've never been able to sleep.

It started, I think, as soon as I learned to read for pleasure. I recall my parents repeatedly coming in my room and telling me to turn off the light. It didn't take a lot of planning to put towels along the cracks at the bottom of the door and over the large old-fashioned keyhole to mask my bedside lamp. One time, when I was reading a story by H. P. Lovecraft, an elegantly eerie horror tale called "The Color Out of Space," I heard a rustling outside my door, looked up, and *saw it!* I believe I said "awk," or something very similar.

The Color Out of Space was a haunted pale blue, and it glowed, in a spectral fashion, as if from within, as if, as if . . .

My eyes adjusted. I could hear my father's heavy tread approaching the door. He must have gotten up to go to the bathroom and turned on the hall light. That light was shining through the keyhole onto the towel I'd hung there, creating the Color Out of Space that had caused me to say "awk," thereby alerting my father to the fact that I was still up at four in the morning. He opened the door, regarded the towels on the floor and the one over the keyhole.

"Go to sleep," my father said. "Right now."

But, of course, I couldn't. He might just as well have told me to go leap mighty buildings in a single bound.

I didn't sleep in high school or college either. I used my insomnia to some economic benefit in graduate school when I had a janitorial job from midnight until eight in the morning, and then attended classes until four in the afternoon.

I don't sleep now that I write for a living. Mostly I write about travel and I haven't slept in some terrific places. I've been awake for twenty-four hours at the North Pole, and for days at a time in Antarctica; I've been sleepless in Seattle, and in Mongolia, and on Tierra del Fuego. It's not so bad. You get to see a lot of nice sunsets, followed by a lot of nice sunrises.

Sleeplessness in your own home, however, is a matter of excruciating agony. Insomnia is not the writer's pal. In my experience, sometime around two in the morning, words become stubborn and disobedient. They sniff each other's crotches, chase their own tails, bark hysterically, steal food off the table, and lift their legs on any standing edifice that represents coherence. Problems often become insurmountable the helpless hours after midnight.

These same problems invariably resolve themselves in as little as fifteen minutes after only a few hours of sleep. All of which means that when I can't sleep, I'm beset with a mountain of insurmountable problems which are most tangibly represented by letters stamped in red with the phrase "past due."

For me, sleep is a trance that channels the muse. My dreams are problem-solving devices. So I lie in bed thinking about the pile of red-stamped letters in my bill box and, in an echo of my father's voice, tell myself to go to sleep, right now. Didn't work then, doesn't work now. The more I worry about not sleeping, the less I sleep.

So, I told my Peruvian friend, what I really wanted to do was go to sleep.

"Ah," he said, "but you must buy for yourself some dreams."

"Buy?"

"*Claro*," the man said. Dreams—*sueños*—were on sale at any pharmacy.

And it came to me that "'dream" was the Peruvian idiom for sleeping pill.

We stopped at the nearest drugstore and purchased a bottle of these dreams. They were medium-sized white pills and the active ingredients listed on the label were unfamiliar.

On the way out of the pharmacy, I asked the man what the fight had been about.

He shrugged. "What else . . . a woman."

"More dreams," I suggested.

"Clearly." He rhapsodized for a time about this woman: how she was indeed like a dream, and how it was his dream to be with her, and how he, like me, had seen fit to fight for his dreams in any way he could.

But now the man had to go home, alone, to an empty apartment, and for this one night, surely, his dream was shattered.

My own night was much the same. A dozen hours later, in the faint ghostly light of false dawn, I heard mourning doves calling outside my hotel window. I'd taken the recommended dose of *sueños* at eleven, and another couple of pills at three. *Nada.* Whatever they put in Peruvian *sueños*, I thought, it's not the stuff dreams are made of.

Siri Hustvedt

M R. MORNING

Sometimes even now I think I see him in the street or standing in a window or bent over a book in a coffee shop. And in that instant, before I understand that it's someone else, my lungs tighten and I lose my breath.

I met him eight years ago. I was a graduate student then at Columbia University. It was hot that summer and my nights were often sleepless. I lay awake in my two-room apartment on West 109th Street listening to the city's noises. I would read, write, and smoke into the morning, but on some nights when the heat made me too listless to work, I watched the neighbors from my bed.

Through my barred window, across the narrow air shaft, I looked into the apartment opposite mine and saw the two men who lived there wander from one room to another, half dressed in the sultry weather. On a day in July, not long before I met Mr. Morning, one of the men came naked to the window. It was dusk and he stood there for a long time, his body lit from behind by a yellow lamp. I hid in the darkness of my bedroom and he never knew I was there. That was two months after Stephen left me, and I thought of him incessantly, stirring in the humid sheets, never comfortable, never relieved.

During the day, I looked for work. In June I had done research for a medical historian. Five days a week I sat in the reading room at the Academy of Medicine on East 102nd Street, filling up index cards with information about great diseases—bubonic plague, leprosy, influenza, syphilis, tuberculosis—as well as more obscure afflictions that I remember now only because of their names—yaws, milk leg, green sickness, rhinitis, ragsorter's disease, housemaid's knee, and dandy fever. Dr. Rosenberg, an octogenarian who spoke and moved very slowly, paid me six dollars an hour for all those index cards, and although I never understood what he did with them, I didn't ask him, fearing that an explanation might take hours. The job ended when my employer went to Italy. I had always been poor as a student, but Dr. Rosenberg's vacation made me desperate. I hadn't paid the July rent, and I had no money for August. Every day I went to the bulletin board in Philosophy Hall where jobs were posted, but by the time I called, they had always been taken. Nevertheless, that was how I found Mr. Morning. A small handwritten notice announced the position: "Wanted. Research assistant for project already under way. Student of literature preferred. Herbert B. Morning." A phone number appeared under the name and I called immediately. Before I could properly introduce myself, a man with a beautiful voice gave me an address on Amsterdam Avenue and told me to come over as soon as possible.

It was hazy that day, but the sun glared and I blinked in the light as I walked through the door of Mr. Morning's tenement building. The elevator was broken, and I remember sweating while I climbed the stairs to the fourth floor. I can still see his intent face in the doorway. He was a very pale man with a large handsome nose. He breathed loudly as he opened the door and let me into a tiny stifling room that smelled of cat. The walls were lined with stuffed bookshelves and more books were piled in leaning towers all over the room. There were tall stacks of newspapers and magazines as well, and beneath a window whose blind had been tightly shut was a heap of old clothes or rags. A massive wooden desk stood in the center of the room and on it were perhaps a dozen boxes of various sizes. Close to the desk was a narrow bed, its rumpled sheets strewn with more books. Mr. Morning seated himself behind the desk, and I sat down in an old folding chair across from him. A narrow ray of light that had escaped through a broken blind fell to the floor between us, and when I looked at it, I saw a haze of dust.

I smoked, contributing to the room's blur, and looked at the skin of his neck; it was moon white. He told me he was happy I had come and then fell silent. Without any apparent reserve, he looked at me, taking in my whole body with his gaze. I don't know if his scrutiny was lecherous or merely curious, but I felt assaulted and turned away from him, and then when he asked me my name, I lied. I did it quickly, without hesitation, inventing a new patronym: Davidsen. I became Iris Davidsen. It was a defensive act, a way of protecting myself from some amorphous danger, but later that false name haunted me; it seemed to move me elsewhere, shifting me off course and strangely altering my whole world for a time. When I think back on it now, I imagine that lie as the beginning of the story, as a kind of door to my uneasiness. Everything else I told him was true—about my parents and sisters in Minnesota, about my studies in nineteenth-century English literature, my past research jobs, even my telephone number. As I talked, he smiled at me, and I thought to myself: It's an intimate smile, as if he has known me for years.

He told me that he was a writer, that he wrote for magazines to earn money. "I write about everything for every taste," he said. "I've written for *Field & Stream, House & Garden, True Confessions, Reader's Digest.* I've written stories, one spy novel, poems, essays, reviews—I even did an art catalogue once." He grinned and waved an arm. "Stanley Rubin's rhythmical canvases reveal a debt to mannerism— Pontormo in particular—the long undulating shapes hint at . . ." He laughed. "And I rarely publish under the same name."

"Don't you stand behind what you write?"

"I am behind everything I write, Miss Davidsen, usually sitting, sometimes standing. In the eighteenth century, it was common to stand and write—at an escritoire. Thomas Wolfe wrote standing."

"That's not exactly what I meant."

"No, of course, it isn't. But you see, Herbert B. Morning couldn't possibly write for *True Confessions,* but Fern Luce can. It's as simple as that."

"You enjoy hiding behind masks?"

"I revel in it. It gives my life a certain color and danger."

"Isn't danger overstating it a bit?"

"I don't think so. Nothing is beyond me as long as I adopt the correct name for each project. It isn't arbitrary. It requires a gift, a genius, if I may say so myself, for hitting on the alias that will unleash the right man or woman for the job. Dewitt L. Parker wrote that art catalogue, for example, and Martin Blane did the spy novel. But there are risks, too. Even the most careful planning can go awry. It's impossible to know for sure who's concealed under the pseudonym I choose."

"I see," I said. "In that case, I should probably ask you who you are now?"

"You have the privilege, dear lady, of addressing Herbert B. Morning as himself, unencumbered by any other personalities."

"And what does Mr. Morning need a research assistant for?"

"For a kind of biography," he said, "for a project about life's paraphernalia, its bits and pieces, treasures and refuse. I need someone like you to respond freely to the objects in question. I need an ear and an eye, a scribe and a voice, a Friday for every day of the week, someone who is sharp, sensitive. You see, I'm in the process of prying open the very essence of the inanimate world. You might say that it's an anthropology of the present."

I asked him to be more specific about the job.

"It began three years ago when she died." He paused as if thinking. "A girl—a young woman. I knew her, but not very well. Anyway, after she died, I found myself in possession of a number of her things, just common everyday things. I had them in the apartment, this and that, out and about, objects that were lost, abandoned, speechless, but not dead. That was the crux of it. They weren't dead, not in the usual way we think of objects as lifeless. They seemed charged with a kind of power. At times I almost felt them move with it, and then after several weeks, I noticed that they seemed to lose that vivacity, seemed to retreat into their thingness. So I boxed them."

"You boxed them?" I said.

"I boxed them to keep them untouched by the here and now. I feel sure that those things carry her imprint—the mark of a warm, living body on the world. And even though I've tried to keep them safe, they're turning cold. I can tell. It's been too long, so my work is urgent. I have to act quickly. I'll pay you sixty dollars per object."

"Per object?" I was sweating in the chair and adjusted my position, pulling my skirt down under my legs, which felt strangely cool to the touch.

"I'll explain everything," he said. He took out a small tape recorder from a drawer in his desk and pushed it toward me. "Listen to this first. It will tell you most of what you want to know. While you listen, I'll leave the room." He stood up from his chair and walked toward a door. A large yellow cat appeared from behind a box and followed him. "Press PLAY," he commanded and vanished.

When I reached for the machine, I noticed two words scrawled on a legal pad near it: "woman's hand." When I turned on the machine, a woman's voice whispered, "This belonged to the deceased. It is a white sheet for a single bed . . ." What followed was a painstaking description of the sheet. It included every tiny discoloration and stain, the texture of the aged cotton and even the tag from which the words had disappeared in repeated washings. It lasted for perhaps ten minutes; the entire speech was delivered in that peculiar half voice. The description itself was tedious and yet I listened with anticipation, imagining that the words would soon reveal something other than the sheet. They didn't. When the tape ended, I looked over to the door behind which Mr. Morning had hidden and saw that it was now ajar and half of his face was pressed through the opening. He was lit from behind, and I couldn't see his features clearly, but the pale hair on his head was shining, and again I heard him breathe with difficulty as he walked toward me. He reached out for my hand. Without thinking, I withdrew it.

"You want descriptions of that girl's things, is that it?" I could hear the tightness and formality in my voice. "I don't understand what a recorded description has to do with your project as a whole or why the woman on the tape was whispering."

"The whisper is essential, because the full human voice is too idiosyncratic, too marked with its own history. I'm looking for ano- nymity so the purity of the object won't be blocked from coming through, from displaying itself in its nakedness. A whisper has no character."

The project seemed odd, even eccentric, but I found myself drawn to it. Chance had given me this small adventure and I was pleased. I also felt that beneath their eccentricity, Mr. Morning's ideas had a weird kind of logic. His comments about whispering, for example, made sense.

"Why don't you write out the descriptions?" I said. "Then there will be no voice at all to interfere with the anonymity you want." I watched his face closely.

He leaned over the desk and looked directly at me. "Because," he said, "then there's no living presence, no force to prompt an awakening."

I shifted in my chair again, gazing at the pile of rags under the window. "What do you mean by awakening?"

"I mean that the objects in question begin to stir under scru- tiny, that they, mute as they are, can nevertheless bear witness to human mysteries."

"You mean they're clues to this girl's life? You want to know about her, is that it? Aren't there more direct routes for finding out biographical information?"

"Not the kind of biography I'm interested in." He smiled at me, this time opening his mouth, and I admired his large white teeth. He isn't old, I thought, not even fifty. He leaned over and picked up a blue box from the floor—a medium-sized department store box—and handed it to me.

I pulled at its lid.

"Not now!" He almost cried out. "Not here."

I pushed the lid back down.

"Do it at home alone. The object must be kept wrapped and in the box unless you are working. Study it. Describe it. Let it speak to you. I have a recorder and a new tape for you as well. Oh yes, and you should begin your description with the words 'This belonged to the deceased.' Could you have it for me by the day after tomorrow?"

I told him I could and then left the apartment with my box and tape recorder, rushing out into the daylight. I walked quickly away from the building and didn't look into the box until I had turned the corner and was sure that he couldn't see me from his window. Inside was a rather dirty white glove lying on a bed of tissue paper.

I didn't go home until later. I fled the heat by going into an air-conditioned coffee shop, sitting for hours as I scribbled notes to myself about the glove and made calculations as to the number of objects I needed to describe before I could pay my rent. I imagined my descriptions as pithy, elegant compositions, small literary exercises based on a kind of belated nineteenth-century positivism. Just for the moment, I decided to pretend that the thing really can be captured by the word. I drank coffee, ate a glazed doughnut, and was happy.

But that night when I put the glove beside my typewriter to begin work, it seemed to have changed. I held it, felt the lumpy wool, and then very slowly pulled it over my left hand. It was too small for my long fingers and didn't cover my wrist. As I looked at it, I had the uncanny feeling that I had seen the same glove on another hand. I began to tug abruptly at its fingers until it sailed to the floor. I let it lie there for several minutes, unwilling to touch it. The small woolen hand covered with smudges and snags seemed terrible to me, a stranded and empty thing, both nonsensical and cruel. Finally I snatched it up and threw it back into the box. There would be no writing until the next day. It was too hot; I was too tired, too nervous. I lay in bed near the open window, but the air stood still. I touched my clammy skin and looked over at the opposite apartment, but the two men had gone to sleep and their windows were black. Before I slept I moved the box into the other room.

That night the screaming began. I woke to the noise but couldn't identify it and thought at first that it was the demented howling of cats I had heard earlier in the summer. But it was a woman's voice—a long guttural wail that ended in a growl. "Stop it! I hate you! I hate you!" she screamed over and over. I stiffened to the noise and wondered if I should call the police, but a long time I just waited and listened. Someone yelled "Shut up" from a window and it stopped. I expected it to begin again, but it was over. I wet a washcloth with cold water and rubbed my neck, arms, and face with it. I thought of Stephen then, as I had often seen him, at his desk, his head turned slightly away from me, his large eyes looking down at a paper. That was when his body was still enchanted; it had a power that I battled and raged against for months. Later that enchantment fell away, and he passed into a banality I never would have thought possible.

The next morning I began again. By daylight the small box on the kitchen table had returned to its former innocence. Using my notes from the coffee shop, I worked steadily, but it was difficult. I looked at the glove closely, trying to remember the words for its various parts, for its texture and the color of its stains. I noticed that the tip of the index finger was blackened, as if the owner had trailed her finger along a filthy surface. She was probably left-handed, I thought, that's a gesture for the favored hand. A girl running her finger along a subway railing. The image prompted a shudder of memory: "woman's hand." The words may have referred to her hand, her gloved hand, or to the glove itself. The connection seemed rife with meaning, and yet it spawned nothing but a feeling vaguely akin to guilt. I pressed on with the description, but the more I wrote, the more specific I was about the glove's characteristics, the more remote it became. Rather than fixing it in the light of scientific exactitude, the abundance of detail made the glove disappear. In fact, my minute description of its discolorations, snags, and pills, its loosened threads and stretched palm, seemed alien to the sad little thing before me.

In the evening I edited my work and then read it into the machine. Whispering bothered me; it made the words clandestine, foreign, and when I listened to the tape, I didn't recognize my own voice. It sounded like a precocious child lisping absurdities from some invisible part of the room, and when I heard it, I blushed with a shame I still don't understand.

Late that night I woke to the screaming again, but it stopped after several minutes, just as before. This time I couldn't get back to sleep and lay awake for hours in a vague torment as the shattered images of exhaustion and heat crowded my brain.

Mr. Morning didn't answer the bell right away. I pressed it three times and was about to leave when I heard him shuffling to the door. He paused in the doorway, looked me directly in the eyes, and smiled. The beautiful smile startled me, and I turned away from him. He apologized for the delay but gave no explanation. That day the apartment seemed more chaotic than on my first visit; the desk in particular was a mass of disturbed papers and boxes. He asked me for the tape; I gave it to him, and then he ushered me from the room, gently pushing me behind the door where he had concealed himself the last time.

I found myself in the kitchen, a tiny room, even hotter and smellier than the other. There were a few unwashed dishes in the sink, several books piled on the counter, and one large white box. From the next room I could just manage to hear the sound of the tape and my soft voice droning on about the glove. I paged through a couple of books, a world atlas and a little copy of *The Cloud of Unknowing*, but I was really interested in the box. I stood over it. The corners of the lid were worn, as if it had been opened many times; two of the sides were taped shut. I ran my finger over the tape to see if I could loosen it. I picked at the tape's yellow skin with my nail, but my efforts made it pucker and tear, so I stopped, trying again on the other side. My head was bent over the box when I heard him coming toward the door, and I leapt backward, accidentally pushing the box off the counter. It fell to the floor but didn't open. I was able to return it to its place before Mr. Morning appeared in the doorway. Whether or not he saw my hands dart away from the box, I still don't know, but when the box fell, whatever was inside it made a loud, hollow, rattling noise, and he must have heard that. Yet he said nothing.

We walked into the other room and sat down. He looked at me and I remember thinking that his gaze had a peculiar strength and that he seemed to blink less often than most people.

"Was the tape O.K.?" I asked.

"Fine," he said, "but there was one aspect of the thing you neglected to describe and I think it's rather important."

"What's that?"

"The odor."

"I didn't think of it," I said.

"No," he said, "many people don't, but without its smell, a thing loses its identity; the absence of odor cripples your description, makes it two-dimensional. Every object has its own scent and carries the odor of its place as well. This can be invaluable to an investigation."

"How?" I said it loudly.

He paused and looked at the window. "By evoking something crucial, something unnoticed before, a place or time or word. Just think of the things we forget in closets and attics, the mildew, the dust, the crushed dry bodies of insects—these odors leave their traces. My mother's trunks smelled of wet wool and lavender. It took me a long time to realize what that odor was, but then I identified it, and I remembered events I had forgotten."

"Is there something you want to remember about this girl who died?" I asked.

"Why do you say that?" He jerked his head toward me.

"Because you obviously want something out of all this. You want these descriptions for a reason. When you mentioned those trunks, I thought you might want to trigger a memory."

He looked away again. "A memory of a whole world," he said.

"But I thought you hardly knew her, Mr. Morning."

He picked up a pencil and began to doodle on a notebook page. "Did I tell you that?"

"Yes, you did."

"It's true. I didn't know her well."

"What is it you're after, then? Who was this person you're investigating?"

"I would like to know that, too."

"You're evading my questions. She had a name, didn't she, this girl?"

"Her name won't help you, Miss Davidsen." His voice was nearly a whisper.

"Well, it won't hurt me either," I said.

He continued to move the pencil idly on the page in front of him. I craned to see it, trying to disguise the gesture by adjusting my skirt. There were several letters written on the paper—what looked like an I, a Y, a B, an O, an M, and a D. He had circled the M. If those markings were intended to form some kind of order, it was impossible to make it out, but even then, before I suspected anything, those letters had a strange effect on me. They stayed with me like the small but persistent aches of a mild illness.

Putting the pencil down, he looked up at me and nodded. He patted his chest. "The heat has given you a rash—here."

"No, it's my birthmark." I touched the skin just below my collarbone.

"A port wine stain," he said. "It has character—a mark for life. If you'll forgive me for saying so, I've always found flaws like that poignant, little outward signs of our mortality. I used a birthmark in something I wrote once . . ."

I interrupted him. "You aren't going to tell me anything, are you?"

"You're referring to our subject, I take it?"

"Of course."

"I think you've failed to understand the nature of your task. I hired you precisely because you know nothing. I hired you to see what I cannot see, because you are who you are. I don't pretend that you're a blank slate. You bring your life with you, your nineteenth-century novels, your Minnesota, the fullness of your existence in every respect, but you didn't know her. When you look at the things I give you, when you write and then speak about them, your words and voice may be the catalysts of some undiscovered being. Knowledge of her will only distract you from your work. Let us say, for the sake of an example, that her name was Allison Hart and that she died of leukemia. Something appears before you, an image. A row of hospital beds, perhaps, in a large room lit by those fluorescent tubes, and you see her, I'm sure you do. Allison—it's a romantic name—pale and emaciated, once beautiful, she lies under white sheets . . . And what you see will not only be shaped by my words, but by my inflections, my expression, and then you will lose your freedom."

I began to speak, but he stopped me.

"No, let me say my piece. Let us say that I tell you her name was"—he paused—"Maxine Robinson and that she was murdered." He looked out past me toward the door and squinted as if he were trying to see something far away. He took several deep breaths.

"That she was killed right here in this building. What would you compose then, Miss Davidsen, when you look inside my boxes? You'd be suffocated by what you know, just as I am. It wouldn't do; it just wouldn't do."

"It doesn't make sense. I still don't understand why I shouldn't know the facts of her life and death."

"Because," he almost shouted. "Because we are about the business of discovery, of resurrection, not burial." He grabbed the edge of the desk and shook it. "Atonement! Miss Davidsen. Atonement!"

"Good God," I said, "atonement for what?"

He was suddenly calm. He pushed his chair back, crossed his legs, folded his arms, and cocked his head to one side. These movements seemed self-conscious, almost theatrical. "For the sins of the world."

"What does that mean?"

"It means exactly what the words say."

"Those words, Mr. Morning," I said, "are liturgical. You've gone into a religious mode all of a sudden. What am I to think? You seem to have a talent for saying nothing with style."

"Be patient and I think you'll begin to understand me." He was smiling.

I had no reply for him. The hot room, the darkness, his outburst, and incomprehensible speech had robbed me of the will to answer. Exhaustion had come over me in a matter of seconds. My bones hurt. Finally I said, "I should leave now."

"If you stay, I'll make tea for you. I'll feed you crumpets and tell you stories. I'll dazzle you with my impeccable manners, my wit and imagination."

I shook my head. "I really have to go."

He paid me then with three twenty-dollar bills and gave me

another box—this time a small white jeweler's box. He told me he didn't need the description until Monday of the following week. I had four days. We shook hands, and then just before I walked through the door, he patted my arm. It was a gesture of sympathy and I accepted it as if it were owed to me.

Inside the second box was a stained and misshapen cotton ball. I found myself hesitant to touch it, as if it were contaminated. The wad of fiber was colored with makeup or powder that looked orange in the light and was also marked with an unidentifiable clot of something dense and brown. I drew away from the little box. Had he salvaged this thing after her death? I imagined him in a bathroom bending over a wastebasket to retrieve the used cotton ball. How had he found these things? Had he hoarded more of her refuse in boxes? I saw him alone, his fingers tracing the outlines of an object as he sat in his chair in front of the window with the closed blinds. But in the daydream I couldn't see what he held. I saw only his body hunched over it.

In those four days between visits to Mr. Morning, I was never free of him. Bits and pieces of his conversation invaded my thoughts, appearing unsummoned at all hours, especially at night. The idea that the girl had been murdered in his building took hold of me, and I began to imagine it. He had taunted me with it; he had intended to entice me with it as just another possible death, but once it was said, I felt that I had known it from the beginning. Resurrection. Atonement. He had seemed genuinely passionate. I remembered his troubled breathing as he spoke, the letters on the page, the white box falling, his hand on my arm. At the same time, I told myself that the man was a charlatan, someone who loved games, riddles, innuendo. Nothing he said could be believed. But in the end, it was his posing that made me suspect that he had hidden the truth among the lies and that he was in earnest about his project and the girl.

That night I worked for hours on the description. I held the cotton ball with a pair of tweezers up to the light, trying to find the words that would express it, but the thing was lost to language; it resisted it even more than the glove. And when I tried metaphors, the object sank so completely into the other thing that I abandoned making comparisons. What was this piece of waste? As I sat sniffing fibers and poking at the brown stain with a needle, I was overwhelmed by a feeling of disgust. The cotton ball told me nothing. It was a blank, a cipher; it probably had no connection to anything terrible, and yet I felt as if I had intruded on a shameful secret, that I had seen what I should not have seen. I composed slowly and my mind wandered. It was a night of many sounds: a man and a woman were fighting in Spanish next door; fire sirens howled and I heard a miserable dog crying somewhere close. I thought of the summers in Minnesota, hot, muggy, but open to the air. At around two o'clock, in the baking confines of my bedroom, I whispered the description into the machine. After it was recorded, I put the cotton ball back into its box and hid it and the tape inside a cupboard in the other room. As I shut the door, I realized that I was behaving like a person with a guilty conscience.

For the third time I stood outside Mr. Morning's door in the dim hallway. A noise was coming from the apartment; it was as if a wind were gusting through it, a rush of sound. I put my ear to the door and then I understood what it was—the tapes, one breathy voice on top of another. He was playing the descriptions. No one voice could be distinguished from another, but I felt sure that mine was among them; I backed away from the door. At that moment I considered running, leaving the box and tape recorder outside the door. Instead I knocked. It may have been that by then I had to know about Mr. Morning, that I had to know what he was hiding. I listened to the sound of the machines being turned off and rewound one by one and then to the sound of drawers being opened and shut.

When he came to the door, he was disheveled. His hair, moist with sweat, stuck up from his head and two buttons on his shirt were unbuttoned. I avoided looking into his flushed face and turned instead to the now familiar room. The blinds were still tightly shut. How can he stand the darkness? I thought. He leaned toward me and smiled.

"Excuse my appearance, Miss Davidsen, I was sleeping and forgot the time altogether. You see me in my Oblomovian persona—only half awake. You'll have to imagine the brocade dressing gown, I'm afraid. And there's no Zakhar, to my infinite regret."

He went on, "Let me have the description and I'll shoo you into the other room right away and then we can talk. I've looked forward to your coming. You brighten the day."

In the kitchen I looked for the box, but it wasn't there. He's moved it, I thought, so I can't see what's in it. The low sound of my voice came from the other room as I waited. How many people had he hired to read those descriptions onto tapes? What were they really for? For an instant, I imagined him lying in the unmade bed listening to that chaos of whispers, but I pushed the image away.

Then he was at the door, motioning for me to follow him into the other room.

"You did a good job with a difficult object," he said.

"Where did you get it?" I said. "It doesn't seem like a very revealing thing to me, a bit of discarded fluff."

"She left it here," he said.

"Who was she? What was she to you?"

"You can't resist, can you? You're dying of curiosity. I suppose it's to be expected from a smart girl like you. I honestly don't know who or what she was to me. If I did, I wouldn't be working on this problem. But that won't satisfy you, will it?"

I heard myself sigh and turned away from him. "I feel that there's something wrong with what you've told me, that there's something hidden behind what you say. It makes me uneasy."

"I will tell you what you want to hear, what you already think you know—that she was murdered. She was killed in the basement laundry room of this building. She lived here."

"And her name was Maxine Robinson."

"No," he said. "I made that up."

"Why?" I said. "Why do that?"

"Because, my friend, I wasn't giving you the facts at the time. I was just giving you a story—one story among a host of possible stories—a little yarn to amuse you and keep you coming back." He looked at his hands. "And keep me alive. A thousand and one tales."

"It would relieve me enormously if you could keep books out of this for once."

"I can try, but they keep popping up like a tic, one after another, rumbling about in my brain, all those people, all that talk. It's a madhouse in there." He pointed to his head and grinned.

"What was her real name?"

"It doesn't matter. I mean that. It doesn't matter for what you're doing. A name can evoke everything and nothing, but it's always a boulder that won't let you pass. I know. I'm a specialist. I want to keep you pure and her nameless." He stared at me. "I'm not fooling you. I need you. I need your help and if you know too much, I'll lose you. You won't be able to do the descriptions anymore."

The emotion in his voice affected me. It was as if he had revealed something intimate, unseemly. I could feel the heat in my face. When I spoke there was a tremor in my voice. "I don't understand you."

"I'm trying to understand a life and an act," he said. "I'm trying to piece together the fragments of an incomprehensible being and to remember. Do you know that I can't even remember her face? Try as I may, it will not be conjured. I can tell you what she looked like; I can recite a description of her features, part by part, but I cannot evoke the whole face."

"Don't you have a photograph?"

"Photographs!" He spat out the word. "I'm talking about true recollection—seeing the face."

The cat rubbed against Mr. Morning's legs and I looked at it. The room was cooler. "Could you open the window?" I said.

He stood up and pulled at the blind, raising it halfway. It was darker outside; a gray cloud cover had replaced the stifling yellow haze. I looked at his profile in front of the window. He stood there in his loose shirt and pants, his hand in one pocket, and I found him elegant. It's in his shoulders, I thought, and the narrowness of his hips. He must have loved her or hated her.

"I should get going," I said.

"You will do another description for me, won't you?"

I nodded. He gave me another small box and three twenty-dollar bills and asked me to return in two days. I pushed the money into my pocket without looking at it and stood up. A breeze came from the window. The weather was changing. At the door, he extended his hand and I took it. He held it for a few seconds longer than he should have, and as I pulled it away, he pressed his thumb into my palm. It startled me, but I felt a familiar shudder of excitement.

It had grown cool with remarkable speed. The sky was darkly overcast and I turned my face upward to feel the first drops of rain as I strode home. I ran into my apartment to open the box, pulling up its lid and pushing aside tissue paper. The third object lay before me on the table. It was a mirror, unadorned, a simple rectangle without even a frame. I picked it up and examined my face, removing a bit of sleep from the inside corner of my eye, studied my mouth, the line of my chin, and then moved the mirror away to see more. I still can't understand it, but as I looked, I was overcome with nausea and faintness. I sat down, put my head between my knees, and took deep breaths. It's possible that the dizziness had nothing to do with the mirror. I had had very little to eat that day and the day before. I scrimped on food for cigarettes, trying to keep my expenses down, and it may have been simple hunger, and yet when I think of that mirror now, it disturbs me, as if there were something wrong with it, something sickening.

Still unstable on my feet, I went to my desk and began to make notes. I was writing to myself, typing out questions about Mr. Morning and the project, but I couldn't put anything together. His remarks about memory, whispering, resurrection returned to me as scraps of some inscrutable idea, some bizarre plan. And then I thought of the noise of the tapes behind the door, his touch and his slender figure in front of the window. Those letters, I thought, those letters on the page. What did they mean? A name. Her name. I moved the letters around, trying to arrange them into a coherent order. I found *mob, boy, dim,* and then *body*. The word coursed through me—a tiny seizure in my nerves. But it was absurd; a man doodles on a paper and I decode his meaningless scribbles. Besides, there were letters that could not be incorporated. *I. M.* He had circled the *M*. The suspicion did not leave me, and I began to imagine that rather than hiding, Mr. Morning really wanted to talk, wanted to tell me something, that the letters, the hints, were revelations, part of a circuitous confession. "If you know too much, I'll lose you." I took my umbrella and went out into the rain.

Within five minutes, I was standing in the entryway of Mr. Morning's building. I buzzed the super. After a considerable wait, a small fat man came to the door. He yawned and then raised his eyebrows, an expression apparently intended to replace the question What do you want?

"I'm looking for an apartment," I said. "Do you have anything vacant?" This was my first ploy, and to my surprise, the building had one empty apartment.

"Three seventy-five a month." He raised his brows again.

"I'd like to see it."

He took me to the third floor and opened the door of a small apartment identical to Mr. Morning's. I walked through the rooms as if I were inspecting them. The man leaned against the open door with a look of belligerent boredom.

"I was told there was a murder in this building," I said.

"That was three long years ago. There hasn't been nothing in that way since."

I walked toward him. "What was her name?"

"Your umbrella's dripping on me, sweetheart."

I moved it away and repeated the question. "Was it Maxine, Maxine Robinson?"

"Hey, hey, hey." He lifted up his hands and backed away from me. "What's going on here? The name was Zalewski, Sherri Zalewski. It's no secret. It was in all the papers."

Tears were in my eyes and I turned away from him.

"What's the matter, kid?" he said.

"Please, tell me," I said. "Did they find the person who did it?"

"You got some kind of special interest here?" he said.

"There can't be any harm in telling me the story," I said.

He did tell me then. I think he was sorry for me or embarrassed by my emotion. Sherri Zalewski had been a nurse who lived in the building. She was knifed to death on a February night while doing her laundry. No one had seen or heard anything. A woman who moved out shortly afterward had found her the next morning. "Real ugly," he said. "Real bad." The woman had vomited in the hallway. The police never found the killer. "They snooped around here for months," he said. "Nothing came of it. They were after the guy in 4F for a while, a real weirdo, Morning. Even took him down to the station. All the tenants were calling and bitching about him. They let him go. Didn't have a thing on him."

"Do you think he killed her?"

"Nah," he said, "he's not the type."

From there I went to Butler Library to check the papers, but there was little new in them. Sherri Zalewski had grown up in Greenpoint. Her mother was dead; her father was a mailman; she had one sister. A friend, quoted in the *Times*, called her "an angel of mercy." Mr. Morning was not mentioned. According to the articles, the police had no suspects. Sherri Zalewski vanished from print for months; her name appeared only once again, in a story run by the *Times* on unsolved murders in New York City. I found a single photograph of her—a grainy block of newsprint that was probably taken from a high school graduation portrait. I stared at the picture, looking for a way in, but it was unusually blank: a girl, neither pretty nor homely, with small eyes and a full mouth.

I carefully attached the chain lock on my door and turned on every light in my apartment before I sat down at the typewriter. I decided to write and record a letter to Mr. Morning. I did describe the mirror briefly but there was little to say. Its surface was unscratched; it had no discernible odor; it was at the same time a full and empty thing, dense with images in one place, vacant in another. Except for the steady sound of the rain outside, my building and street were uncommonly quiet that night, but the noises I did hear made me jump, and I understood that I was listening for someone, waiting, expecting the sound of an intruder. He was in my head. Fragments of our conversation came back to me: Fern Luce, what he had said about remembering the girl's face, the smell of wool and lavender in his mother's trunk. I wrote, and as I wrote, I saw her body on the floor in the vacant apartment I had visited. I always see it there, for some reason—bloodied and torn apart. I see the corpse as in a photograph, in black and white, illuminated by a dim light bulb. Even now when it comes to me, I can't examine it closely. I push it away.

Evening became night. The room turned dusky and a chill made the blond hair on my arms stand up. I wrapped myself in a blanket and wrote one page after another and threw them away.

When I finished I had just one page. The mirror lay beside me shining in the lamplight. At around one o'clock in the morning, I spoke the words I had written into the tape recorder but didn't listen to them. The wind blew over my bed, and I fell into a deep, empty sleep.

Mr. Morning's rooms were cool and wet that day. His windows were open for the first time, and the wind blew in, ruffling a newspaper that lay on top of a pile. His usually pale cheeks were rosy and he seemed to be breathing more easily. I am quite sure that he sensed my apprehension immediately, because he said so little to me, and in his face there was sorrow or maybe regret. Before I secluded myself in the kitchen, I noticed that there was a tall stack of papers that looked like a manuscript on the desk.

I didn't close the door to the kitchen; I let it stand open slightly and put my eye to the crack. I watched him as he placed the tape recorder in front of him on the desk and turned it on. He leaned back in his chair, let his arms hang limply at his sides, and closed his eyes. After a brief interval of static from the machine, I heard my voice come from the other room. I listened to the short description of the mirror that I had dutifully whispered onto the tape. Then I heard my full voice and saw Mr. Morning look sharply in my direction. I quickly shut the door. As I listened to the high, childlike voice that must have been mine, I clenched my teeth so tightly that later my jaw was sore.

"I know who she was. Her name was Sherri Zalewski. I wondered for a while if you hadn't invented her, but now I know that she existed and that she lived and died in your building. A glove, a stained cotton ball, a mirror. Why these things? Where did you find them? You must have known that I would ask these questions. I suspect that you have invited them, that you knew I would find out about her and about you. You should have told me the story, Mr. Morning. You should have told me directly rather than hinting at it. I do believe that, for you, this project is somehow an attempt to undo what happened that night, that these things of hers are a part of some elaborate idea that I can't make out." There was a pause on the tape, and I listened for a noise from him, but there was nothing. "The things, the tapes, all your talk. I don't know what to do with them, how to understand them, how to understand you. I do know that the dead do not come back to life." I heard a loud scraping noise. He must have moved his chair. But the tape was still on. I pressed myself against the door as if the weight of my body could shut him out. "I know that the police questioned you, that they suspected you. I am not saying that you killed her; I'm asking you to tell me the truth. That is all." It was over. He was walking to the door and I heard him turn the knob on the other side. I stepped back. He was breathing loudly and a wheezing sound seemed to come from deep in his chest. He stood in the open doorway and stared at me, his face deeply flushed. He looked as if he were about to speak, but then he closed his mouth and gained control of his breathing.

He said, "What is there to say? You expect me to confess, don't you, to fall down before you and tell you that I murdered her. But that isn't going to happen. It can't happen."

"What are you saying?" My voice was choked.

"I have already explained everything to you." He looked past me and pressed his lips together in a spasm of emotion. "There is nothing more to say. The story is yours. Not mine."

"What do you mean?"

"I mean that you've invented the story yourself. It belongs to you, not to me. You've already chosen an ending, a way out. I suppose it's inevitable that you should want satisfaction." He looked at me. " 'The evil wizard turned to stone.' 'The King and Queen lived happily ever after.' 'He died in prison, a broken man.' Whatever. What you've forgotten is that some things are unspeakable. That's what you've left out. Words may cover it up for a while, but then it comes howling back. A storm. A plague. Only half remembered. The difference between you and me is that I know I've forgotten. You don't." He turned around and faced the other room.

I spoke to his back. "That's what you have to say to me? I ask you to tell me the truth and you tell me that?"

"Yes," he said.

"I don't understand you. I don't understand you at all. Tell me that you didn't kill her." My voice was shrill.

"No," he said.

Mr. Morning walked toward his desk, and I heard the blinds rattle. There was a gust of wind from outside, and the papers on the desk were whipped into the air—hundreds of white pages flapped noisily against the bookshelves and walls, blew over the chairs and stacks of newspapers, sliding across the wood floor. Mr. Morning scrambled to retrieve them.

"Listen, Iris," he said. "I know things have changed, but I don't want to lose you. I want you to stay with me and do some more work. I want you to talk with me the way you've done these last two weeks. You will stay, won't you?"

I said yes to him. I thought to myself that if I did one more description, I could press him again, that he would tell me the truth, but now I wonder if that was really the reason.

He opened the desk drawer and took out another small white box. He held it out to me with both hands. "For tomorrow. Tomorrow at two." He gave me the tape recorder, and then, after explaining that he was short of cash, he wrote out a check to Iris Davidsen.

"I can't accept it," I said.

"Please, I insist."

I took it, knowing that I could never cash it. I walked to the door, picking my way among the fallen pages. He walked beside me.

At the door, he took my hand in both of his. "There's one last thing. Before you go, I want you to leave me something of yours." His eyes were shining.

"No."

"Why not?"

I released my hand from his grip. "No."

"One small thing." He leaned closer to me, and in the opening of his shirt I saw the cleft of his collarbone. There was a vague scent of cologne.

I opened my purse and began to search it, roughly pushing aside books, envelopes, and keys until I found an old green eraser, blackened with lead smudges, and thrust it into his hand, saying that I had to leave for an appointment.

I imagine that he stood in the doorway and watched me rush to the stairs and that he continued to stand there as I ran down one flight after another, because I never heard the door close.

I ran into the street and began to walk toward Broadway. When I reached the corner, I paused. It had stopped raining and the sky was breaking into vast, blank holes of blue. I watched the clouds move and then looked into the street. The sidewalk, buildings, and people had been given a fierce clarity in the new light; each thing appeared radically distinct, as if my eyesight had suddenly been sharpened. It was then that I decided to get rid of the things. I opened my bag, took out the check, ripped it to pieces, and threw it into a large trash bin. Then I threw away the tape recorder and the unopened box. I can still see the small black machine lying askew on the garbage heap and the smaller box as it tumbled farther into the bin. It upset a Styrofoam cup as it fell, and I turned away just as a stream of pale brown coffee dregs ran over its lid. My memory of those discarded objects, lying among the other waste, is vivid but silent, as if I had been standing in the noiseless city of a movie or a dream. I saw them for only an instant, and then I ran from those things as if they were about to rise up and pursue me.

I didn't think that would be the end of it. Mr. Morning had my telephone number, after all, and there was nothing to prevent him from finding me. I waited for months, but I never heard from him. When the telephone rang, it was always someone else.

E. Annie Proulx

WAKING UP

Any damn old thing wakes me up—the howling of wolves, thunder, the flyswatter on a nail swinging in the night breeze, a fax from Australia, twisted bedclothes, a sojourner spider, nothing at all.

I don't think of it as insomnia when I wake up at 11:00 after half an hour of sleep, or at 1:30 or 4 A.M. Wake-ups are common in my nights, the times I read and do bookish research, or, if the hour is creeping toward dawn, drink a cup of hot water, read an old copy of the *Times*—it takes a week to get a copy of *The New York Times* by mail out in Wyoming and usually they clog the mailbox in age-yellowed batches of three and four, so I always have a supply of stale news. Plenty of books, too, stacked around the bed in case I don't want to get up—a score within pillow reach just now, including Gerald Lynch's *Roughnecks, Drillers, and Tool-Pushers*, Dermot Healy's *A Goat's Song*, already dog-eared from one intense reading, Karal Ann Marling's *As Seen on TV*, Kahn and Whitehead's *Wireless Imagination*, John Stilgoe's *Shallow Water Dictionary*, the 1941 WPA guide to Wyoming, Byrne's *Standard Book of Pool and Billiards*. There's something for every hour of the night and plenty more on the other side of the room.

In normal sleepless hours I work, scribbling away on yellow pads, but for the past month my right wrist has been immobilized in a cast and my labored attempts with the pen look like ransom notes.

Unlike the virtuous, I never bake bread in the middle of the night or clean house or weed out the Rolodex. I might put on an assortment of Quebec reels and waltzes and leap and clog for half an hour as well as one can in bunny slippers. Sometimes I make speeches, sing loudly and horribly *Hawhawhohum*, read aloud Dakota poet Byron B. Bobb's "The Carnage of Badaxe," fill out order blanks in junk mail catalogues, cross bridges that I have not yet come to, make coffee and drink it.

Waking up in the night is no problem if you live alone (no one is disturbed by your blaze of light and rustling pages) and work at home (who cares if you fall asleep with your face on the table?). Long ago I stopped thinking of wakefulness as an anomaly.

There is the dark side of the insomniac's adventures and that is the languorous drowsiness that o'ertakes one at noon, the eye-swimming need to sleep, the staggering somnambulist's gait, and, finally, the surrender.

I'm at that point now.

I'm going back to bed.

F . Scott Fitzgerald

SLEEPING

AND WAKING

December 1934

When some years ago I read a piece by Ernest Hemingway called "Now I Lay Me," I thought there was nothing further to be said about insomnia. I see now that that was because I had never had much; it appears that every man's insomnia is as different from his neighbor's as are their daytime hopes and aspirations.

Now if insomnia is going to be one of your naturals, it begins to appear in the late thirties. Those seven precious hours of sleep suddenly break in two. There is, if one is lucky, the "first sweet sleep of night" and the last deep sleep of morning, but between the two appears a sinister, ever widening interval. This is the time of which it is written in the Psalms: *Scuto circumdabit te veritas eius: non timebis a timore nocturno, a sagitta volante in die, a negotio perambulante in tenebris.*

With a man I knew the trouble commenced with a mouse; in my case I like to trace it to a single mosquito.

My friend was in course of opening up his country house unassisted, and after a fatiguing day discovered that the only practical bed was a child's affair—long enough but scarcely wider than a crib. Into this he flopped and was presently deeply engrossed in rest *but* with one arm irrepressibly extending over the side of the crib. Hours later he was awakened by what seemed to be a pin-prick in his finger. He shifted his arm sleepily and dozed off again—to be again awakened by the same feeling.

This time he flipped on the bed-light—and there attached to the bleeding end of his finger was a small and avid mouse. My friend, to use his own words, "uttered an exclamation," but probably he gave a wild scream.

The mouse let go. It had been about the business of devouring the man as thoroughly as if his sleep were permanent. From then on it threatened to be not even temporary. The victim sat shivering, and very, very tired. He considered how he would have a cage made to fit over the bed and sleep under it the rest of his life. But it was too late to have the cage made that night and finally he dozed, to wake in intermittent horrors from dreams of being a Pied Piper whose rats turned about and pursued him.

He has never since been able to sleep without a dog or cat in the room.

My own experience with night pests was at a time of utter exhaustion—too much work undertaken, interlocking circumstances that made the work twice as arduous, illness within and around—the old story of troubles never coming singly. And ah, how I had planned that sleep that was to crown the end of the struggle—how I had looked forward to the relaxation into a bed soft as a cloud and permanent as a grave. An invitation to dine *à deux* with Greta Garbo would have left me indifferent.

But had there been such an invitation I would have done well to accept it, for instead I dined alone, or rather was dined upon by one solitary mosquito.

It is astonishing how much worse one mosquito can be than a swarm. A swarm can be prepared against, but *one* mosquito takes on a personality—a hatefulness, a sinister quality of the struggle to the death. This personality appeared all by himself in September on the twentieth floor of a New York hotel, as out of place as an armadillo. He was the result of New Jersey's decreased appropriation for swamp drainage, which had sent him and other younger sons into neighboring states for food.

The night was warm—but after the first encounter, the vague slappings of the air, the futile searches, the punishment of my own ears a split second too late, I followed the ancient formula and drew the sheet over my head.

And so there continued the old story, the bitings through the sheet, the sniping of exposed sections of hand holding the sheet in place, the pulling up of the blanket with ensuing suffocation—followed by the psychological change of attitude, increasing wakefulness, wild impotent anger—finally a second hunt.

This inaugurated the maniacal phase—the crawl under the bed with the standing lamp for torch, the tour of the room with final detection of the insect's retreat on the ceiling and attack with knotted towels, the wounding of oneself—my God!

—After that there was a short convalescence that my opponent seemed aware of, for he perched insolently beside my head—
but I missed again.

At last, after another half hour that whipped the nerves into a
frantic state of alertness came the Pyrrhic victory, and the small
mangled spot of blood, *my* blood, on the head-board of the bed.

As I said, I think of that night, two years ago, as the beginning
of my sleeplessness—because it gave me the sense of how sleep can
be spoiled by one infinitesimal incalculable element. It made me, in
the now archaic phraseology, "sleep-conscious." I worried whether
or not it was going to be allowed me. I was drinking, intermittently
but generously, and on the nights when I took no liquor the problem of whether or not sleep was specified began to haunt me long
before bedtime.

A typical night (and I wish I could say such nights were all in
the past) comes after a particularly sedentary work-and-cigarette
day. It ends, say without any relaxing interval, at the time for going
to bed. All is prepared, the books, the glass of water, the extra
pajamas lest I awake in rivulets of sweat, the Luminal pills in the
little round tube, the note book and pencil in case of a night
thought worth recording. (Few have been—they generally seem
thin in the morning, which does not diminish their force and urgency at night.)

I turn in, perhaps with a night-cap—I am doing some compar-
atively scholarly reading for a coincident work so I choose a lighter
volume on the subject and read till drowsy on a last cigarette. At the
yawning point I snap the book on a marker, the cigarette at the
hearth, the button on the lamp. I turn first on the left side, for that,
so I've heard, slows the heart, and then—coma.

So far so good. From midnight until two-thirty peace in the

room. Then suddenly I am awake, harassed by one of the ills or functions of the body, a too vivid dream, a change in the weather for warm or cold.

The adjustment is made quickly, with the vain hope that the continuity of sleep can be preserved, but no—so with a sigh I flip on the light, take a minute pill of Luminal and reopen my book. The *real* night, the darkest hour, has begun. I am too tired to read unless I get myself a drink and hence feel bad next day—so I get up and walk. I walk from my bedroom through the hall to my study, and then back again, and if it's summer out to my back porch. There is a mist over Baltimore; I cannot count a single steeple. Once more to the study, where my eye is caught by a pile of unfinished business: letters, proofs, notes, etc. I start toward it, but No! this would be fatal. Now the Luminal is having some slight effect, so I try bed again, this time half circling the pillow on edge about my neck.

"Once upon a time" (I tell myself) "they needed a quarterback at Princeton, and they had nobody and were in despair. The head coach noticed me kicking and passing on the side of the field, and he cried: 'Who is *that* man—why haven't we noticed *him* before?' The under coach answered, 'He hasn't been out,' and the response was: 'Bring him to me.'

". . . we go to the day of the Yale game. I weigh only one hundred and thirty-five, so they save me until the third quarter, with the score—"

—But it's no use—I have used that dream of a defeated dream to induce sleep for almost twenty years, but it has worn thin at last. I can no longer count on it—though even now on easier nights it has a certain lull . . .

The war dream then: The Japanese are everywhere victorious—my division is cut to rags and stands on the defensive in a part of Minnesota where I know every bit of the ground. The headquarters staff and the regimental battalion commanders who were in conference with them at the time have been killed by one shell. The command devolved upon Captain Fitzgerald. With superb presence . . .

—But enough; this also is worn thin with years of usage. The character who bears my name has become blurred. In the dead of the night I am only one of the dark millions riding forward in black buses toward the unknown.

Back again now to the rear porch, and conditioned by intense fatigue of mind and perverse alertness of the nervous system—like a broken-stringed bow upon a throbbing fiddle—I see the real horror develop over the roof-tops, and in the strident horns of night-owl taxis and the shrill monody of revelers' arrival over the way. Horror and waste—

—Waste and horror—what I might have been and done that is lost, spent, gone, dissipated, unrecapturable. I could have acted thus, refrained from this, been bold where I was timid, cautious where I was rash.

I need not have hurt her like that.

Nor said this to him.

Nor broken myself trying to break what was unbreakable.

The horror has come now like a storm—what if this night prefigured the night after death—what if all thereafter was an eternal quivering on the edge of an abyss, with everything base and vicious in oneself urging one forward and the baseness and viciousness of the world just ahead. No choice, no road, no hope—only the endless repetition of the sordid and the semi-tragic. Or to stand forever, perhaps, on the threshold of life unable to pass it and return to it. I am a ghost now as the clock strikes four.

On the side of the bed I put my head in my hands. Then silence, silence—and suddenly—or so it seems in retrospect—suddenly I am asleep.

Sleep—real sleep, the dear, the cherished one, the lullaby. So deep and warm the bed and the pillow enfolding me, letting me sink into peace, nothingness—my dreams now, after the catharsis of the dark hours, are of young and lovely people doing young, lovely things, the girls I knew once with big brown eyes, real yellow hair.

> *In the fall of '16 in the cool of the afternoon*
> *I met Caroline under a white moon*
> *There was an orchestra—Bingo-Bango*
> *Playing for us to dance the tango*
> *And the people all clapped as we arose*
> *For her sweet face and my new clothes—*

Life *was* like that, after all; my spirit soars in the moment of its oblivion; then down, down deep into the pillow . . .

". . . Yes, Essie, yes. —Oh, My God, all right, I'll take the call myself."

Irresistible, iridescent—here is Aurora—here is another day.

George Dawes Green

THE CHAINS
OF CIRCADIA

Cop rolls into the Brewster diner at four in the morning and checks out the clientele, the usual congeries of wheezers and greasers, the slow dunkers and the gassy drooling gaffers. Up at the counter there's an old guy they call WATCH-OUT!, who's advising the ether that everything is due for a CHANGE! Everything's gonna be UPENDED! Nobody's listening to WATCH-OUT! except one little boy, maybe nine years old, sitting two stools down and drinking a Coke and happily absorbing Mr. WATCH-OUT!'s teleological alarums.

The cop eases himself onto the stool next to the kid. "Where you from, kid?"

"Carmel."

"Jesus! How'd you get here?"

"I walked."

"On Route 6? All the way from Carmel?"

"I walked on the old railroad tracks."

"In the middle of the night? Come on, don't lie to me. It's *pitch black* out there."

"There's lots of stars out."

This colloquy at the Brewster diner, this happened a long time ago, and I presume that since then WATCH-OUT! has been radiantly upended—just as he prophesied—translated into that very star-quilt that the cop hadn't even *noticed.*

The cop says, "You had a fight with your folks?"

"No." The boy smiles. "I just felt like taking a walk."

"Aren't you tired?"

"No."

"Don't you want to go sleep?"

"I'm not tired. Are you going to arrest me, Officer?"

"Nah. Just gonna take you on home. Jesus."

Few weeks later another cop picked me up a little north of Ludingtonville at dawn.

I'd walked twenty miles, starting a little after midnight. I was on my way to the Catskills, because I'd always wanted to see the mountains. The cop told me it would have taken me a week to walk to the Catskills.

The cop said, "Don't you think you could learn to sleep during the night, like the rest of us?"

"I like to look at the stars."

Also I rapturously charted for him all the phases of sunrise I'd seen—from the first faltering of the eastern constellations to the tentative early thrushes by the Kent firehouse, to the great trumpet-blast breakthrough of the sun over a gas station by Ludingtonville Pond.

The cop asked me if my school had a school psychiatrist.

He took me home and my father cooked me breakfast. My father told me if I was going to wander the earth like Marley's ghost I'd better not get caught any more, because next time the police were going to take me to a home for delinquents. I went up to my bed at ten-thirty in the morning. The day was blooming outside my window, and I couldn't sleep for bliss. When finally I did drop off, I slept for eleven hours—and by midnight I was back on the railroad tracks.

That was summer. That was my life in the forgiving summer-time.

Come the start of school though, I was back in the Bondage of 24.

Monday morning at seven o'clock my mother would drag my corpse out of bed. Set it upright. Give it a shove toward the school bus, and I'd snooze on the way to school, and I'd drift in and out of homeroom and math and social studies. When I came home I'd nap some more but as soon as my parents were in bed I'd sneak downstairs and watch Steve Allen or *The Joe Pyne Show*. The guests on *The Joe Pyne Show* looked like the denizens of the Brewster diner. They were quacks and nut cases. I watched close up to the TV screen, with the sound turned down low. Joe's guests held elaborate theories about reincarnation, they brought with them tables showing their alien genealogies. They would ramble on till poor impatient

Joe could suffer them no more, then he would jerk his thumb and off they would shuffle, pursued by catcalls. But I believed them all.

Then it was eight o'clock in the morning again and I was sleepwalking through school.

There was something the matter with me.

My day didn't seem to *fit* inside the sun's day. The clock and I were at odds.

I did poorly in school. They called me an underachiever.

When I was old enough, my parents sent me to a prep school in Massachusetts, thinking the discipline would be good for me.

But prep school couldn't fit me into the Bondage of 24 either. I started sleeping through breakfast. Then through lunch. One day I finally *did* make it to breakfast, and my tablemates rose and applauded. I took a bow. Then I went back to my room and slept through lunch and dinner, and at two in the morning I arose refreshed and ravenous and walked to a diner outside of Springfield. All my old friends from *The Joe Pyne Show* were there. WATCH-OUT! was there. Everyone smoked cigarettes and stared into his coffee cup. We believed that Alpha Centaurian blood oozed in our veins, we believed that we hailed from noble planets with generous, easygoing rotations and fat orbits. We considered the Bondage of 24, the Chains of Circadia, and shook our heads slowly. We were all busy underachieving, we were cursed, outcast of men, and dwelling in the margins. I was among my brothers and sisters, and we had nothing to say to each other, and for a few moments I was content.

A year or so later I dropped out of school and went into the working world. In the working world the Bondage of 24 has full dominion. Satan's alarm clocks are always exploding in your ear, jitterbugging on your bedside table. I was miserable. I could not wake up when I was supposed to; I could not sleep when I was supposed to.

How can you get ahead in the world when you're sitting at an employment interview and your eyelids are the weight of naval cannons, and you're holding them up with grappling hooks and turning blue from the effort? How can you hope to advance yourself? All the man wants is your Social Security number, but you're afraid if you lift your tongue to answer you'll have to first let go of those grappling hooks and then your eyelids will drop down, down into that sweet black abyss, that black, briny, sweet . . .

"Sir? Are you *asleep?*"

A thousand scolds, rousing me from my slumbers.

"What the hell is the matter with you?"

"What the hell kind of schedule are you *on*, son?"

I didn't know. Now I do.

After many years of that harrowing nightmare, after many weird careers and regimens attempted, after so much sleep-deprived, befogged suffering—I finally managed to save some money, and I decided to take some time off and write a novel.

And, more to the point, I determined that I would keep my own schedule for a while.

I got heavy curtains for my bedroom. I banished all clocks, telephones, TVs. I got a fan and draped a towel over it and left it close to my ear so that while I slept I would hear no report of the outside world, nothing to disturb me.

I would sleep when I pleased and wake up when I pleased and to hell with 24.

. . .

And I discovered that, left alone, I live a 26-hour day.

I like to stay up for seventeen hours and sleep nine. One morning I'll rise at five in the morning. You'll think I'm a go-getter, Ben Franklin's boy; a healthy, birdwatching, cow-milking Child of Nature. The next day I'll wake up at seven and you'll think I'm just like you. The next day I'll sleep till nine and you'll smile indulgently.

Then I'll get up at eleven and you'll call me lazy. Then I'll sleep till one in the afternoon, then three, then *five*—and you'll call me a sybaritic clod, you'll stand outside my door and denounce me, in the name of Phoebus you'll rain curses on my sweet slumber—but I'll have that fan going full blast by my left ear, and I won't hear you.

By and by my schedule will roll around till I'm waking at nine in the evening. I'll be up all night and you're just a little scared of me now, aren't you? Aren't you slightly concerned for the safety of your nine-year-old kid when ghouls like me are prowling the streets? You send cops around to pick me up, to round us all up. But the all-night diners are our sanctuary, in the diners 24 cannot touch us.

Here and there a few of us are slipping the Chains of Circadia, the word is spreading, I tell you that surely the doom of the Bondage of 24 is nigh. Soon outside your door a restless legion of Alpha Centaurians will be gathering, and we're proud of our far-flung heritage and WATCH-OUT! will be among us. Armed with our heavy curtains and our towel-draped fans, we'll all soon be *overachieving like crazy*, O my brothers and sisters. I have beheld our true ancestral day, and it's as loose-limbed and languorous as we remembered, so follow me. Smash the clocks and spit on their tyrannical grinding innards, march a million strong onto the cramped studio stage of *The Joe Pyne Show*, and when he jerks his thumb at us? BITE IT OFF!

We like to look at the stars, Officer. Are you going to arrest all of us?

Jonathan Carroll

C RIMES

OF THE FACE

My father was a careful man. He taught us to count our change before leaving a store, to check the tires on our bikes before taking them out of the garage, to brush our teeth in an up-and-down rather than side-to-side motion.

Because he was careful and because he handled other people's money as if it were his own, he became successful and wealthy and we lived well. He was the man you see raking leaves in front of the nice Connecticut house on an orange-and-brown fall day. Or the one at the A&P with a couple of his kids picking up supplies for the barbecue—ten pounds of briquettes, corn on the cob, a couple of steaks as thick as telephone books.

The only thing that was odd about him was he was cursed with increasingly bad insomnia as he got older. One of my vivid memories of childhood is waking up in the middle of the night and going to the bathroom or downstairs for something to eat. He was always up, either reading in the living room or standing at the kitchen counter eating an egg-and-onion sandwich, which, for some peculiar reason, he said helped him to sleep. He was always happy to see me, and now with an adult's understanding, I'm sure it was because he was glad for the company even if it was only for a few minutes before I went sleepily back up the stairs to bed.

I say this because among the other things I inherited from him, the insomnia is unfortunately one of them. He used to say he was lucky to have it because he got to own more of the day than anyone else and loved the mysterious quiet of the deep night. I do not, but that is not important to this story, which, in an important way, hinges on one sleepless night I experienced not long ago.

I think he had a lover once but even now cannot be sure. I would never in a million years ask my mother, however, because she loved him very much and since he died has spent the years sifting her memories of him through some kind of benevolent strainer that leaves her with only happy or sadly-sweet memories of their life together.

The reason why I mention this lover is because of my father's one real peculiarity: he was an absolute dandy when it came to clothes. He spent thousands and thousands of dollars on them and was never satisfied with what he had. Never. Gray suits, blue suits, gray-blue suits. A tie rack that took up one whole door of his custom-made closet. A closet that was so off-limits to us kids that even the thought of going near it gave us the shivers. When I learned German in school, I realized the only word that properly applied to my father's closet was VERBOTEN. "Forbidden" has too many soft, sissy sounds in it.

He went to Europe once a year on business for a few weeks. A month or two after he returned, big beautifully wrapped boxes from places like Charvet or Hilditch and Key would arrive at the door, full of silk ties or cotton shirts white and thick as milk. A suit? He bought suits, plural. He knew half the salesmen at Paul Stuart and I'm sure had one of the first charge cards from Brooks Brothers.

Clothes didn't mean anything to us except now and then we absolutely had to have the exact same sneakers as Bob Cousy or Willie Naulls wore on the basketball court. Other than that, if our dad owned a million sweaters, so what? So what if Mom's closet was half the size of his, or that she regularly tsk'd her tongue whenever he came into the house with another shoe box under his arm, looking vaguely naughty but at the same time hugely pleased. If I were to ask her about this today, my mother would smile and say, "His only vice—new suits. Give him a new suit and he was happy for a month."

But it's not that simple. I have my own children now and what worries me most about them is what they will remember twenty years from now that I wish like hell they'd forget. I know from my own experience children remember nothing logical or momentous so much as things quirky and appropriate to their personalities. When I was nine I saw a man run over right in front of me, but that's just a lot of vague, unpleasant images now. What I really remember down to the last detail these many years later was going to my first New York Yankees baseball game with Dad and, because it was cold, keeping my hands in his overcoat pocket the whole time. That was also the first time I ever tasted coffee, because he said it would warm me up. I can still feel the warmth of it on the edge of the paper cup.

It's the old adage about giving a child a present: it doesn't matter how wonderful or expensive it is, because if the child is very young, they'll like the box more than the present. Give them the biggest Christmas ever, but when they're grown they'll remember only how hard the marzipan was in their Christmas stocking that morning.

Normally my father was a calm, solid fellow. But when he was preparing for his trip to Europe, he raced around the house all in a flurry for two or three days before he left. All of us tried to stay out of his way as much as possible because he was like a train that's gone too fast and is almost out of control. I used to think his flurrying was because he was harried and edgy about his trip, but now I'm not so sure. What I remember most distinctly was the collective sigh of relief we'd all breathe when he was finally out the door and we had the house to ourselves again. Not that we wanted him to go—just that we wanted that *part* of him gone. We knew when he returned he'd be our old dad and pal again and things would be well until next year's trip.

I am the only one of the men in our family who looks like my father. It is startling how over the years I have come to resemble him more and more. At forty-five I have the same kind of wrinkles, same slight smile, the same head of peppery brown hair that he had until the day he died. Not to mention the insomnia. And I'm glad of that, because I liked my father very much. I liked his manner and the way he thought about the world, the way he dealt with his often rowdy household. But now I think I know something that makes me like him even more. It is not the sort of thing you are supposed to either know or like about your father. Yet I do and it makes me smile. The secret sharer. However, I am not one hundred percent sure that it is the truth or if there is even a glimmer of the truth in it, but I like to think so.

Several months ago I was in Vienna for a medical conference. I don't like to travel and even the lure of a piano recital at the Konzerthaus there did little to wipe the frown off my lips when I picked up my bag and headed out the front door of the house. So unlike my father in this, I approach any trip like others do battle.

I was staying at the Sacher Hotel, as you are supposed to do if you're only going to be in Vienna once in your life. Unfortunately, with the combination of the insomnia and jet lag, I found myself up half the nights I was there either prowling the streets or finding myself at whatever open bars I could find in that quiet town with an unwanted drink in front of me and a desperate longing to close my eyes and rest for a few merciful hours.

Halfway through the conference I found myself sitting at the Sacher bar very late one night drinking alone and wishing for the thousandth time I could either go upstairs to sleep or go home.

With my back to the door, I only saw people when they were all the way into the room. Since I wasn't watching too carefully, I didn't see or hear her until she had said his name and put her small hand on my shoulder.

I have left out something very important. When my father died four years ago at a very contented seventy-nine, my mother asked if I would like his wardrobe. My brothers thought I was out of my mind (or at least ghoulish) when I said yes. I felt it both an honor and a gift of great import to own and continue to wear such beautiful things. I knew I would appreciate them for the rest of my life and, psychiatrists be damned, wearing my father's clothes made me feel like I had been given some of the great style I had always admired him for as I grew older and could appreciate it.

I was wearing one of his suits that night in the Sacher bar. A favorite gray flannel of his that was cut just so that he always took with him when he went to Europe, because, Mother said, he felt like the king of the world in it.

I felt a hand on my shoulder, and a woman's voice, tentative and sweet, said my father's name the way the French pronounce it: "On-ree?"

The voice both asked a question and made a kind of frightened statement at the same time. I turned, not knowing what to expect. What I saw was a woman of about sixty-five, striking and stately and apparently from some Eastern European country, because when she spoke again it was with a very thick accent. Also her face had those eerily high cheekbones and deep-set eyes one sees in photographs of Russian gentry at the turn of the century.

When I turned and faced her completely, that marvelous ruined face set into a frightened, almost stunned expression. She was seeing a ghost and she could feel the material of his familiar jacket under her fingertips. Face to face with a part of her past that had suddenly become the very real present with no reason why.

Whatever the case may be, she said his name again, a statement, the whole time staring at me as hard as a human being can stare. Then she said in a dry whisper, "But it is *impossible!* You must be more than seventy now. More! Like me! It was forty years ago!"

Her hand was gripping my shoulder now, her fingers digging in. She looked at it there and then, at the wonderful suit that made my father feel like a king. Recognition flooded her face, but it was horrendous, impossible stuff.

Before I had a chance to say anything, she snatched her hand away and, shaking her head violently from side to side, fled the room.

Had her "On-ree" been my father? Had she touched his suit, *this* suit, with love and expectation as they rode in black boxy taxis across London, coming from the theater, going to dinner, the great magical moments of both their lives here now, this evening, however long he could stay with her this time. Later the suit and her clothes on the floor together of a hotel room at Brown's or the Connaught. Two figures out of the movies we watch now on the Late Show, telling ourselves, "God, I wish life was like that!"

I can't be sure. It is impossible to tell. But what I have been thinking about recently is something more terrible, particularly when I am awake again at three-thirty in the morning and clearer about things than I sometimes like. If she did believe it was my father that night, or even some other Henry she had known and loved long ago, then all of the rites and rituals of the world she had gone by for all of her long life were suddenly and irrevocably wrong. The kind of sacred mysteries we must avoid because they are monstrous. Mysteries, like lost sleep, that leave us hopelessly alone in the middle of the night with too many hours of shadow and silence to bear.

Haruki Murakami

S L E E P

This is my seventeenth straight day without sleep.

I'm not talking about insomnia. I know what insomnia is. I had
something like it in college—"something like it" because I'm not
sure that what I had then was exactly the same as what people refer
to as insomnia. I suppose a doctor could have told me. But I didn't
see a doctor. I knew it wouldn't do any good. Not that I had any
reason to think so. Call it woman's intuition—I just felt they
couldn't help me. So I didn't see a doctor, and I didn't say anything
to my parents or friends, because I knew that that was exactly what
they would tell me to do.

Back then, my "something like insomnia" went on for a month. I never really got to sleep that entire time. I'd go to bed at night and say to myself, "All right now, time for some sleep." That was all it took to wake me up. It was instantaneous—like a conditioned reflex. The harder I worked at sleeping, the wider awake I became. I tried alcohol, I tried sleeping pills, but they had absolutely no effect.

Finally, as the sky began to grow light in the morning, I'd feel that I might be drifting off. But that wasn't sleep. My fingertips were just barely brushing against the outermost edge of sleep. And all the while, my mind was wide awake. I would feel a hint of drowsiness, but my mind was there, in its own room, on the other side of a transparent wall, watching me. My physical self was drifting through the feeble morning light, and all the while it could feel my mind staring, breathing, close beside it. I was both a body on the verge of sleep and a mind determined to stay awake.

This incomplete drowsiness would continue on and off all day. My head was always foggy. I couldn't get an accurate fix on the things around me—their distance or mass or texture. The drowsiness would overtake me at regular, wavelike intervals: on the subway, in the classroom, at the dinner table. My mind would slip away from my body. The world would sway soundlessly. I would drop things. My pencil or my purse or my fork would clatter to the floor. All I wanted was to throw myself down and sleep. But I couldn't. The wakefulness was always there beside me. I could feel its chilling shadow. It was the shadow of myself. Weird, I would think as the drowsiness overtook me, I'm in my own shadow. I would walk and eat and talk to people inside my drowsiness. And the strangest thing was that no one noticed. I lost fifteen pounds that month, and no one noticed. No one in my family, not one of my friends or classmates, realized that I was going through life asleep.

It was literally true: I was going through life asleep. My body had no more feeling than a drowned corpse. My very existence, my life in the world, seemed like a hallucination. A strong wind would make me think that my body was about to be blown to the end of the earth, to some land I had never seen or heard of, where my mind and body would separate forever. Hold tight, I would tell myself, but there was nothing for me to hold on to.

And then, when night came, the intense wakefulness would return. I was powerless to resist it. I was locked in its core by an enormous force. All I could do was stay awake until morning, eyes wide open in the dark. I couldn't even think. As I lay there, listening to the clock tick off the seconds, I did nothing but stare at the darkness as it slowly deepened and slowly diminished.

And then one day it ended, without warning, without any external cause. I started to lose consciousness at the breakfast table. I stood up without saying anything. I may have knocked something off the table. I think someone spoke to me. But I can't be sure. I staggered to my room, crawled into bed in my clothes, and fell fast asleep. I stayed that way for twenty-seven hours. My mother became alarmed and tried to shake me out of it. She actually slapped my cheeks. But I went on sleeping for twenty-seven hours without a break. And when I finally did awaken, I was my old self again. Probably.

I have no idea why I became an insomniac then or why the condition suddenly cured itself. It was like a thick, black cloud brought from somewhere by the wind, a cloud crammed full of ominous things I have no knowledge of. No one knows where such a thing comes from or where it goes. I can only be sure that it did descend on me for a time, and then departed.

. . .

In any case, what I have now is nothing like that insomnia, nothing at all. I just can't sleep. Not for one second. Aside from that simple fact, I'm perfectly normal. I don't feel sleepy, and my mind is as clear as ever. Clearer, if anything. Physically, too, I'm normal: My appetite is fine; I'm not fatigued. In terms of everyday reality, there's nothing wrong with me. I just can't sleep.

Neither my husband nor my son has noticed that I'm not sleeping. And I haven't mentioned it to them. I don't want to be told to see a doctor. I know it wouldn't do any good. I just know. Like before. This is something I have to deal with myself.

So they don't suspect a thing. On the surface, our life flows on unchanged. Peaceful. Routine. After I see my husband and son off in the morning, I take my car and go marketing. My husband is a dentist. His office is a ten-minute drive from our condo. He and a dental-school friend own it as partners. That way, they can afford to hire a technician and a receptionist. One partner can take the other's overflow. Both of them are good, so for an office that has been in operation for only five years and that opened without any special connections, the place is doing very well. Almost too well. "I didn't want to work so hard," says my husband. "But I can't complain."

And I always say, "Really, you can't." It's true. We had to get an enormous bank loan to open the place. A dental office requires a huge investment in equipment. And the competition is fierce. Patients don't start pouring in the minute you open your doors. Lots of dental clinics have failed for lack of patients.

Back then, we were young and poor and we had a brand-new

baby. No one could guarantee that we would survive in such a tough world. But we have survived, one way or another. Five years. No, we really can't complain. We've still got almost two thirds of our debt left to pay back, though.

"I know why you've got so many patients," I always say to him. "It's because you're such a good-looking guy."

This is our little joke. He's not good-looking at all. Actually, he's kind of strange-looking. Even now I sometimes wonder why I married such a strange-looking man. I had other boyfriends who were far more handsome.

What makes his face so strange? I can't really say. It's not a handsome face, but it's not ugly, either. Nor is it the kind that people would say has "character." Honestly, "strange" is about all that fits. Or maybe it would be more accurate to say that it has no distinguishing features. Still, there must be some element that *makes* his face have no distinguishing features, and if I could grasp whatever that is, I might be able to understand the strangeness of the whole. I once tried to draw his picture, but I couldn't do it. I couldn't remember what he looked like. I sat there holding the pencil over the paper and couldn't make a mark. I was flabbergasted. How can you live with a man so long and not be able to bring his face to mind? I knew how to recognize him, of course. I would even get mental images of him now and then. But when it came to drawing his picture, I realized that I didn't remember anything about his face. What could I do? It was like running into an invisible wall. The one thing I could remember was that his face looked strange.

The memory of that often makes me nervous.

Still, he's one of those men everybody likes. That's a big plus in his business, obviously, but I think he would have been a success at just about anything. People feel secure talking to him. I had never met anyone like that before. All my women friends like him. And I'm fond of him, of course. I think I even love him. But strictly speaking, I don't actually *like* him.

Anyhow, he smiles in this natural, innocent way, just like a child. Not many grown-up men can do that. And I guess you'd expect a dentist to have nice teeth, which he does.

"It's not my fault I'm so good-looking," he always answers when we enjoy our little joke. We're the only ones who understand what it means. It's a recognition of reality—of the fact that we have managed in one way or another to survive—and it's an important ritual for us.

He drives his Sentra out of the condo parking garage every morning at 8:15. Our son is in the seat next to him. The elementary school is on the way to the office. "Be careful," I say. "Don't worry," he answers. Always the same little dialogue. I can't help myself. I have to say it. "Be careful." And my husband has to answer, "Don't worry." He starts the engine, puts a Haydn or a Mozart tape into the car stereo, and hums along with the music. My two "men" always wave to me on the way out. Their hands move in exactly the same way. Their hands move in exactly the same way. It's almost uncanny. They lean their heads at exactly the same angle and turn their palms toward me, moving them slightly from side to side in exactly the same way, as if they'd been trained by a choreographer.

I have my own car, a used Honda Civic. A girlfriend sold it to me two years ago for next to nothing. One bumper is smashed in, and the body style is old-fashioned, with rust spots showing up. The odometer has over 150,000 kilometers on it. Sometimes—once or twice a month—the car is almost impossible to start. The engine simply won't catch. Still, it's not bad enough to have the thing fixed. If you baby it and let it rest for ten minutes or so, the engine will start up with a nice, solid *vroom*. Oh, well, everything—every-body—gets out of whack once or twice a month. That's life. My husband calls my car "your donkey." I don't care. It's mine.

I drive my Civic to the supermarket. After marketing, I clean the house and do the laundry. Then I fix lunch. I make a point of performing my morning chores with brisk, efficient movements. If possible, I like to finish my dinner preparations in the morning, too. Then the afternoon is all mine.

My husband comes home for lunch. He doesn't like to eat out. He says the restaurants are too crowded, the food is no good, and the smell of tobacco smoke gets into his clothes. He prefers eating at home, even with the extra travel time involved. Still, I don't make anything fancy for lunch. I warm up leftovers in the microwave or boil a pot of noodles. So the actual time involved is minimal. And of course it's more fun to eat with my husband than all alone with no one to talk to.

Before, when the clinic was just getting started, there would often be no patient in the first afternoon slot, so the two of us would go to bed after lunch. Those were the loveliest times with him. Everything was hushed, and the soft afternoon sunshine would filter into the room. We were a lot younger then, and happier.

We're still happy, of course. I really do think so. No domestic troubles cast shadows on our home. I love him and trust him. And I'm sure he feels the same about me. But little by little, as the months and years go by, your life changes. That's just how it is. There's nothing you can do about it. Now all the afternoon slots are taken. When we finish eating, my husband brushes his teeth, hurries out to his car, and goes back to the office. He's got all those sick teeth waiting for him. But that's all right. We both know you can't have everything your own way.

After my husband goes back to the office, I take a bathing suit and towel and drive to the neighborhood athletic club. I swim for half an hour. I swim hard. I'm not that crazy about the swimming itself: I just want to keep the flab off. I've always liked my own figure. Actually, I've never liked my face. It's not bad, but I've never really liked it. My body is another matter. I like to stand naked in front of the mirror. I like to study the soft outlines I see there, the balanced vitality. I'm not sure what it is, but I get the feeling that something inside there is very important to me. Whatever it is, I don't want to lose it.

I'm thirty. When you reach thirty, you realize it's not the end of the world. I'm not especially happy about getting older, but it does make some things easier. It's a question of attitude. One thing I know for sure, though: If a thirty-year-old woman loves her body and is serious about keeping it looking the way it should, she has to put in a certain amount of effort. I learned that from my mother. She used to be a slim, lovely woman, but not anymore. I don't want the same thing to happen to me.

After I've had my swim, I use the rest of my afternoon in various ways. Sometimes I'll wander over to the station plaza and window-shop. Sometimes I'll go home, curl up on the sofa, and read a book or listen to the FM station or just rest. Eventually, my son comes home from school. I help him change into his playclothes, and give him a snack. When he's through eating, he goes out to play with his friends. He's too young to go to an afternoon cram school, and we aren't making him take piano lessons or anything. "Let him play," says my husband. "Let him grow up naturally."

When my son leaves the house, I have the same little dialogue with him as I do with my husband. "Be careful," I say, and he answers, "Don't worry."

As evening approaches, I begin preparing dinner. My son is always back by six. He watches cartoons on TV. If no emergency patients show up, my husband is home before seven. He doesn't drink a drop and he's not fond of pointless socializing. He almost always comes straight home from work.

The three of us talk during dinner, mostly about what we've done that day. My son always has the most to say. Everything that happens in his life is fresh and full of mystery. He talks, and we offer our comments. After dinner, he does what he likes—watches television or reads or plays some kind of game with my husband. When he has homework, he shuts himself up in his room and does it. He goes to bed at 8:30. I tuck him in and stroke his hair and say good night to him and turn off the light.

Then it's husband and wife together. He sits on the sofa, reading the newspaper and talking to me now and then about his patients or something in the paper. Then he listens to Haydn or Mozart. I don't mind listening to music, but I can never seem to tell the difference between those two composers. They sound the same to me. When I say that to my husband, he tells me it doesn't matter. "It's all beautiful. That's what counts."

"Just like you," I say.

"Just like me," he answers with a big smile. He seems genuinely pleased.

. . .

So that's my life—or my life before I stopped sleeping—each day pretty much a repetition of the one before. I used to keep a diary, but if I forgot for two or three days, I'd lose track of what had happened on which day. Yesterday could have been the day before yesterday, or vice versa. I'd sometimes wonder what kind of life this was. Which is not to say that I found it empty. I was—very simply—amazed. At the lack of demarcation between the days. At the fact that I was part of such a life, a life that had swallowed me up so completely. At the fact that my footprints were being blown away before I even had a chance to turn and look at them.

Whenever I felt like that, I would look at my face in the bathroom mirror—just look at it for fifteen minutes at a time, my mind a total blank. I'd stare at my face purely as a physical object, and gradually it would disconnect from the rest of me, becoming just some thing that happened to exist at the same time as myself. And a realization would come to me: This is happening here and now. It's got nothing to do with footprints. Reality and I exist simultaneously at this present moment. That's the most important thing.

But now I can't sleep anymore. When I stopped sleeping, I stopped keeping a diary.

I remember with perfect clarity that first night I lost the ability to sleep. I was having a repulsive dream—a dark, slimy dream. I don't remember what it was about, but I do remember how it felt: ominous and terrifying. I woke at the climactic moment—came fully awake with a start, as if something had dragged me back at the last moment from a fatal turning point. Had I remained immersed in the dream for another second, I would have been lost forever. After I awoke, my breath came in painful gasps for a time. My arms and legs felt paralyzed. I lay there immobilized, listening to my own labored breathing, as if I were stretched out full-length on the floor of a huge cavern.

"It was a dream," I told myself, and I waited for my breathing to calm down. Lying stiff on my back, I felt my heart working violently, my lungs hurrying the blood to it with big, slow, bellow-slike contractions. I began to wonder what time it could be. I wanted to look at the clock by my pillow, but I couldn't turn my head far enough. Just then, I seemed to catch a glimpse of something at the foot of the bed, something like a vague, black shadow. I caught my breath. My heart, my lungs, everything inside me, seemed to freeze in that instant. I strained to see the black shadow.

The moment I tried to focus on it, the shadow began to assume a definite shape, as if it had been waiting for me to notice it. Its outline became distinct, and began to be filled with substance, and then with details. It was a gaunt old man wearing a skintight black shirt. His hair was gray and short, his cheeks sunken. He stood at my feet, perfectly still. He said nothing, but his piercing eyes stared at me. They were huge eyes, and I could see the red network of veins in them. The old man's face wore no expression at all. It told me nothing. It was like an opening in the darkness.

This was no longer the dream, I knew. From that I had already awakened. And not just by drifting awake, but by having my eyes ripped open. No, this was no dream. This was reality. And in reality an old man I had never seen before was standing at the foot of my bed. I had to do something—turn on the light, wake my husband, scream. I tried to move. I fought to make my limbs work, but it did no good. I couldn't move a finger. When it became clear to me that

I would never be able to move, I was filled with a hopeless terror, a primal fear such as I had never experienced before, like a chill that rises silently from the bottomless well of memory. I tried to scream, but I was incapable of producing a sound or even moving my tongue. All I could do was look at the old man.

Now I saw that he was holding something—a tall, narrow, rounded thing that shone white. As I stared at this object, wondering what it could be, it began to take on a definite shape, just as the shadow had earlier. It was a pitcher, an old-fashioned porcelain pitcher. After some time, the man raised the pitcher and began pouring water from it onto my feet. I could not feel the water. I could see it and hear it splashing down onto my feet, but I couldn't feel a thing.

The old man went on and on pouring water over my feet. Strange—no matter how much he poured, the pitcher never ran dry. I began to worry that my feet would eventually rot and melt away. Yes, of course they would rot. What else could they do with so much water pouring over them? When it occurred to me that my feet were going to rot and melt away, I couldn't take it any longer.

I closed my eyes and let out a scream so loud it took every ounce of strength I had. But it never left my body. It reverberated soundlessly inside, tearing through me, shutting down my heart. Everything inside my head turned white for a moment as the scream penetrated my every cell. Something inside me died. Something melted away, leaving only a shuddering vacuum. An explosive flash incinerated everything my existence depended on.

When I opened my eyes, the old man was gone. The pitcher was gone. The bedspread was dry, and there was no indication that anything near my feet had been wet. My body, though, was soaked with sweat, a horrifying volume of sweat, more sweat than I ever imagined a human being could produce. And yet, undeniably, it was sweat that had come from me.

I moved one finger. Then another, and another, and the rest. Next, I bent my arms and then my legs. I rotated my feet and bent my knees. Nothing moved quite as it should have, but at least it did move. After carefully checking to see that all my body parts were working, I eased myself into a sitting position. In the dim light filtering in from the streetlamp, I scanned the entire room from corner to corner. The old man was definitely not there.

The clock by my pillow said 12:30. I had been sleeping for only an hour and a half. My husband was sound asleep in his bed. Even his breathing was inaudible. He always sleeps like that, as if all mental activity in him had been obliterated. Almost nothing can wake him.

I got out of bed and went into the bathroom. I threw my sweat-soaked nightgown into the washing machine and took a shower. After putting on a fresh pair of pajamas, I went to the living room, switched on the floor lamp beside the sofa, and sat there drinking a full glass of brandy. I almost never drink. Not that I have a physical incompatibility with alcohol, as my husband does. In fact, I used to drink quite a lot, but after marrying him I simply stopped. Sometimes when I had trouble sleeping I would take a sip of brandy, but that night I felt I wanted a whole glass to quiet my overwrought nerves.

The only alcohol in the house was a bottle of Rémy-Martin we kept in the sideboard. It had been a gift. I don't even remember who gave it to us, it was so long ago. The bottle wore a thin layer of dust. We had no real brandy glasses, so I just poured it into a regular tumbler and sipped it slowly.

I must have been in a trance, I thought. I had never experienced such a thing, but I had heard about trances from a college friend who had been through one. Everything was incredibly clear, she had said. You can't believe it's a dream. "I didn't believe it was a dream when it was happening, and now I still don't be- lieve it was a dream." Which is exactly how I felt. Of course it had to be a dream—a kind of dream that doesn't feel like a dream.

Though the terror was leaving me, the trembling of my body would not stop. It was in my skin, like the circular ripples on water after an earthquake. I could see the slight quivering. The scream had done it. That scream that had never found a voice was still locked up in my body, making it tremble.

I closed my eyes and swallowed another mouthful of brandy. The warmth spread from my throat to my stomach. The sensation felt tremendously *real*.

With a start, I thought of my son. Again my heart began pounding. I hurried from the sofa to his room. He was sound asleep, one hand across his mouth, the other thrust out to the side, looking just as secure and peaceful in sleep as my husband. I straightened his blanket. Whatever it was that had so violently shattered my sleep, it had attacked only me. Neither of them had felt a thing.

I returned to the living room and wandered about there. I was not the least bit sleepy.

I considered drinking another glass of brandy. In fact, I wanted to drink even more alcohol than that. I wanted to warm my body more, to calm my nerves down more, and to feel that strong, penetrating bouquet in my mouth again. After some hesitation, I decided against it. I didn't want to start the new day drunk. I put the brandy back in the sideboard, brought the glass to the kitchen sink and washed it. I found some strawberries in the refrigerator and ate them.

I realized that the trembling in my skin was almost gone.

What was that old man in black? I asked myself. I had never seen him before in my life. That black clothing of his was so strange, like a tight-fitting sweat suit, and yet, at the same time, old-fashioned. I had never seen anything like it. And those eyes— bloodshot, and never blinking. Who was he? Why did he pour water onto my feet? Why did he have to do such a thing?

I had only questions, no answers.

The time my friend went into a trance, she was spending the night at her fiancé's house. As she lay in bed asleep, an angry-looking man in his early fifties approached and ordered her out of the house. While that was happening, she couldn't move a muscle. And, like me, she became soaked with sweat. She was certain it must be the ghost of her fiancé's father, who was telling her to get out of his house. But when she asked to see a photograph of the father the next day, it turned out to be an entirely different man. "I must have been feeling tense," she concluded. "That's what caused it."

But *I'm* not tense. And this is my own house. There shouldn't be anything here to threaten me. Why did *I* have to go into a trance?

I shook my head. Stop thinking, I told myself. It won't do any good. I had a realistic dream, nothing more. I've probably been building up some kind of fatigue. The tennis I played the day before yesterday must have done it. I met a friend at the club after my swim and she invited me to play tennis and I overdid it a little, that's all. Sure—my arms and legs felt tired and heavy for a while afterward.

When I finished my strawberries, I stretched out on the sofa and tried closing my eyes.

I wasn't sleepy at all. Oh, great, I thought. I really don't feel like sleeping.

I thought I'd read a book until I got tired again. I went to the bedroom and picked a novel from the bookcase. My husband didn't even twitch when I turned on the light to hunt for it. I chose *Anna Karenina*. I was in the mood for a long Russian novel, and I had read *Anna Karenina* only once, long ago, probably in high school. I remembered just a few things about it: the first line, "All happy families resemble one another; every unhappy family is unhappy in its own way," and the heroine's throwing herself under a train at the end. And that early on there was a hint of the final suicide. Wasn't there a scene at a racetrack? Or was that in another novel?

Whatever. I went back to the sofa and opened the book. How many years had it been since I'd sat down and relaxed like this with a book? True, I often spent half an hour or an hour of my private time in the afternoon with a book open. But you couldn't really call that reading. I'd always find myself thinking about other things— my son, or shopping, or the freezer's needing to be fixed, or my having to find something to wear to a relative's wedding, or the stomach operation my father had last month. That kind of stuff would drift into my mind, and then it would grow and take off in a million different directions. After a while I'd notice that the only thing that had gone by was the time, and I had hardly turned any pages.

Without noticing it, I had become accustomed in this way to a life without books. How strange, now that I think of it. Reading had been the center of my life when I was young. I had read every book in the grade-school library, and almost my entire allowance would go for books. I'd even scrimp on lunches to buy books I wanted to read. And this went on into junior high and high school. Nobody read as much as I did. I was the third of five children, and both my parents worked, so nobody paid much attention to me. I could read alone as much as I liked. I'd always enter the essay contests on books so that I could win a gift certificate for more books. And I usually won. In college, I majored in English literature and got good grades. My graduation thesis on Katherine Mansfield won top honors, and my thesis adviser urged me to apply to graduate school. I wanted to go out into the world, though, and I knew that I was no scholar. I just enjoyed reading books. And even if I had wanted to go on studying, my family didn't have the financial wherewithal to send me to graduate school. We weren't poor by any means, but there were two sisters coming along after me, so once I graduated from college I simply had to begin supporting myself.

When had I really read a book last? And what had it been? I couldn't recall anything. Why did a person's life have to change so completely? Where had the old me gone, the one who used to read a book as if possessed by it? What had those days—and that almost abnormally intense passion—meant to me?

That night, I found myself capable of reading *Anna Karenina* with unbroken concentration. I went on turning pages without another thought in mind. In one sitting, I read as far as the scene where Anna and Vronsky first see each other in the Moscow train station. At that point, I stuck my bookmark in and poured myself another glass of brandy.

Though it hadn't occurred to me before, I couldn't help thinking what an odd novel this was. You don't see the heroine, Anna, until Chapter 18. I wondered if it didn't seem unusual to readers in

Tolstoy's day. What did they do when the book went on and on with a detailed description of the life of a minor character named Oblonsky—just sit there, waiting for the beautiful heroine to appear? Maybe that was it. Maybe people in those days had lots of time to kill—at least the part of society that read novels.

Then I noticed how late it was. Three in the morning! And still I wasn't sleepy.

What should I do? I don't feel sleepy at all, I thought. I could just keep on reading. I'd love to find out what happens in the story. But I have to sleep.

I remembered my ordeal with insomnia and how I had gone through each day back then, wrapped in a cloud. No, never again. I was still a student in those days. It was still possible for me to get away with something like that. But not now, I thought. Now I'm a wife. A mother. I have responsibilities. I have to make my husband's lunches and take care of my son.

But even if I get into bed now, I know I won't be able to sleep a wink.

I shook my head.

Let's face it, I'm just not sleepy, I told myself. And I want to read the rest of the book.

I sighed and stole a glance at the big volume lying on the table. And that was that. I plunged into *Anna Karenina* and kept reading until the sun came up. Anna and Vronsky stared at each other at the ball and fell into their doomed love. Anna went to pieces when Vronsky's horse fell at the racetrack (so there *was* a racetrack scene, after all!) and confessed her infidelity to her husband. I was there with Vronsky when he spurred his horse over the obstacles. I heard the crowd cheering him on. And I was there in the stands watching his horse go down. When the window brightened with the morning light, I laid down the book and went to the kitchen for a cup of coffee. My mind was filled with scenes from the novel and with a tremendous hunger obliterating any other thoughts. I cut two slices of bread, spread them with butter and mustard, and had a cheese sandwich. My hunger pangs were almost unbearable. It was rare for me to feel that hungry. I had trouble breathing, I was so hungry. One sandwich did hardly anything for me, so I made another one and had another cup of coffee with it.

To my husband I said nothing about either my trance or my night without sleep. Not that I was hiding them from him. It just seemed to me that there was no point in telling him. What good would it have done? And besides, I had simply missed a night's sleep. That much happens to everyone now and then.

I made my husband his usual cup of coffee and gave my son a glass of warm milk. My husband ate toast, and my son ate a bowl of cornflakes. My husband skimmed the morning paper, and my son hummed a new song he had learned in school. The two of them got into the Sentra and left. "Be careful," I said to my husband. "Don't worry," he answered. The two of them waved. A typical morning.

After they were gone, I sat on the sofa and thought about how to spend the rest of the day. What should I do? What did I have to do? I went to the kitchen to inspect the contents of the refrigerator. I could get by without shopping. We had bread, milk, and eggs, and there was meat in the freezer. Plenty of vegetables, too. Everything I'd need through tomorrow's lunch.

I had business at the bank, but it was nothing I absolutely had to take care of immediately. Letting it go a day longer wouldn't hurt.

I went back to the sofa and started reading the rest of *Anna Karenina*. Until that reading, I hadn't realized how little I remembered of what goes on in the book. I recognized virtually nothing—the characters, the scenes, nothing. I might as well have been reading a whole new book. How strange. I must have been deeply moved at the time I first read it, but now there was nothing left. Without my noticing, the memories of all the shuddering, soaring emotions had slipped away and vanished.

What, then, of the enormous fund of time I had consumed back then reading books? What had all that meant?

I stopped reading and thought about that for a while. None of it made sense to me, though, and soon I even lost track of what I was thinking about. I caught myself staring at the tree that stood outside the window. I shook my head and went back to the book.

Just after the middle of Volume 3, I found a few crumbling flakes of chocolate stuck between the pages. I must have been eating chocolate as I read the novel when I was in high school. I used to like to eat and read. Come to think of it, I hadn't touched chocolate since my marriage. My husband doesn't like me to eat sweets, and we almost never give them to our son. We don't usually keep that kind of thing around the house.

As I looked at the whitened flakes of chocolate from over a decade ago, I felt a tremendous urge to have the real thing. I wanted to eat chocolate while reading *Anna Karenina*, the way I did back then. I couldn't bear to be denied it for another moment. Every cell in my body seemed to be panting with this hunger for chocolate.

I slipped a cardigan over my shoulders and took the elevator down. I walked straight to the neighborhood candy shop and bought two of the sweetest-looking milk-chocolate bars they had. As soon as I left the shop, I tore one open and started eating it while walking home. The luscious taste of milk chocolate spread through my mouth. I could feel the sweetness being absorbed directly into every part of my body. I continued eating in the elevator, steeping myself in the wonderful aroma that filled the tiny space.

Heading straight for the sofa, I started reading *Anna Karenina* and eating my chocolate. I wasn't the least bit sleepy. I felt no physical fatigue, either. I could have gone on reading forever. When I finished the first chocolate bar, I opened the second and ate half of that. About two thirds of the way through Volume 3, I looked at my watch. Eleven-forty.

Eleven-forty!

My husband would be home soon. I closed the book and hurried to the kitchen. I put water in a pot and turned on the gas. Then I minced some scallions and took out a handful of buckwheat noodles for boiling. While the water was heating, I soaked some dried seaweed, cut it up, and topped it with a vinegar dressing. I took a block of tofu from the refrigerator and cut it into cubes. Finally, I went into the bathroom and brushed my teeth to get rid of the chocolate smell.

At almost the exact moment the water came to a boil, my husband walked in. He had finished work a little earlier than usual, he said.

Together, we ate the buckwheat noodles. My husband talked about a new piece of dental equipment he was considering bringing into the office, a machine that would remove plaque from patients' teeth far more thoroughly than anything he had used before, and in less time. Like all such equipment, it was quite expensive, but it would pay for itself soon enough. More and more patients were coming in just for a cleaning these days.

"What do you think?" he asked me.

I didn't want to think about plaque on people's teeth, and I especially didn't want to hear or think about it while I was eating. My mind was filled with hazy images of Vronsky falling off his horse. But of course I couldn't tell my husband that. He was deadly serious about the equipment. I asked him the price and pretended to think about it. "Why not buy it if you need it?" I said. "The money will work out one way or another. You wouldn't be spending it for fun, after all."

"That's true," he said. "I wouldn't be spending it for fun." Then he continued eating his noodles in silence.

Perched on a branch of the tree outside the window, a pair of large birds was chirping. I watched them half-consciously. I wasn't sleepy. I wasn't the least bit sleepy. Why not?

While I cleared the table, my husband sat on the sofa reading the paper. *Anna Karenina* lay there beside him, but he didn't seem to notice. He had no interest in whether I read books.

After I finished washing the dishes, my husband said, "I've got a nice surprise today. What do you think it is?"

"I don't know," I said.

"My first afternoon patient has canceled. I don't have to be back in the office until one-thirty." He smiled.

I couldn't figure out why this was supposed to be such a nice surprise. I wonder why I couldn't.

It was only after my husband stood up and drew me toward the bedroom that I realized what he had in mind. I wasn't in the mood for it at all. I didn't understand why I should have sex then. All I wanted was to get back to my book. I wanted to stretch out alone on the sofa and munch on chocolate while I turned the pages of *Anna Karenina*. All the time I had been washing the dishes, my only thoughts had been of Vronsky and of how an author like Tolstoy managed to control his characters so skillfully. He described them with wonderful precision. But that very precision somehow denied them a kind of salvation. And this finally—

I closed my eyes and pressed my fingertips to my temple.

"I'm sorry, I've had a kind of headache all day. What awful timing."

I had often had some truly terrible headaches, so he accepted my explanation without a murmur.

"You'd better lie down and get some rest," he said. "You've been working too hard."

"It's really not that bad," I said.

He relaxed on the sofa until one o'clock, listening to music and reading the paper. And he talked about dental equipment again. You bought the latest high-tech stuff and it was obsolete in two or three years. . . . So then you had to keep replacing everything. . . . The only ones who made any money were the equipment manufacturers—that kind of talk. I offered a few clucks, but I was hardly listening.

After my husband went back to the office, I folded the paper and pounded the sofa cushions until they were puffed up again. Then I leaned on the windowsill, surveying the room. I couldn't figure out what was happening. Why wasn't I sleepy? In the old days, I had done all-nighters any number of times, but I had never stayed awake this long. Ordinarily, I would have been sound asleep after so many hours or, if not asleep, impossibly tired. But I wasn't the least bit sleepy. My mind was perfectly clear.

I went into the kitchen and warmed up some coffee. I

thought, Now what should I do? Of course, I wanted to read the rest of *Anna Karenina*, but I also wanted to go to the pool for my swim. I decided to go swimming. I don't know how to explain this, but I wanted to purge my body of something by exercising it to the limit. Purge it—of what? I spent some time wondering about that. Purge it of what?

I didn't know.

But this thing, whatever it was, this mistlike something, hung there inside my body like a certain kind of potential. I wanted to give it a name, but the word refused to come to mind. I'm terrible at finding the right words for things. I'm sure Tolstoy would have been able to come up with exactly the right word.

Anyhow, I put my swimsuit in my bag and, as always, drove my Civic to the athletic club. There were only two other people in the pool—a young man and a middle-aged woman—and I didn't know either of them. A bored-looking lifeguard was on duty.

I changed into my bathing suit, put on my goggles, and swam my usual thirty minutes. But thirty minutes wasn't enough. I swam another fifteen minutes, ending with a crawl at maximum speed for two full lengths. I was out of breath, but I still felt nothing but energy welling up inside my body. The others were staring at me when I left the pool.

It was still a little before three o'clock, so I drove to the bank and finished my business there. I considered doing some shopping at the supermarket, but I decided instead to head straight for home. There, I picked up *Anna Karenina* where I had left off, eating what was left of the chocolate. When my son came home at four o'clock, I gave him a glass of juice and some fruit gelatin that I had made. Then I started on dinner. I defrosted some meat from the freezer and cut up some vegetables in preparation for stir-frying. I made miso soup and cooked the rice. All of these tasks I took care of with tremendous mechanical efficiency.

I went back to *Anna Karenina*.

I was not tired.

At ten o'clock, I got into my bed, pretending that I would be sleeping there near my husband. He fell asleep right away, practically the moment the light went out, as if there were some cord connecting the lamp with his brain.

Amazing. People like that are rare. There are far more people who have trouble falling asleep. My father was one of those. He'd always complain about how shallow his sleep was. Not only did he find it hard to get to sleep, but the slightest sound or movement would wake him up for the rest of the night.

Not my husband, though. Once he was asleep, nothing could wake him until morning. We were still newlyweds when it struck me how odd this was. I even experimented to see what it would take to wake him. I sprinkled water on his face and tickled his nose with a brush—that kind of thing. I never once got him to wake up. If I kept at it, I could get him to groan once, but that was all. And he never dreamed. At least he never remembered what his dreams were about. Needless to say, he never went into any paralytic trances. He slept. He slept like a turtle buried in mud.

Amazing. But it helped with what quickly became my nightly routine.

After ten minutes of lying near him, I would get out of bed. I would go to the living room, turn on the floor lamp, and pour myself a glass of brandy. Then I would sit on the sofa and read my book, taking tiny sips of brandy and letting the smooth liquid glide over my tongue. Whenever I felt like it, I would eat a cookie or a piece of chocolate that I had hidden in the sideboard. After a while, morning would come. When that happened, I would close my book and make myself a cup of coffee. Then I would make a sandwich and eat it.

My days became just as regulated.

I would hurry through my housework and spend the rest of the morning reading. Just before noon, I would put my book down and fix my husband's lunch. When he left, before one, I'd drive to the club and have my swim. I would swim for a full hour. Once I stopped sleeping, thirty minutes was never enough. While I was in the water, I concentrated my entire mind on swimming. I thought about nothing but how to move my body most effectively, and I inhaled and exhaled with perfect regularity. If I met someone I knew, I hardly said a word—just the basic civilities. I refused all invitations. "Sorry," I'd say. "I'm going straight home today. There's something I have to do." I didn't want to get involved with anybody. I didn't want to have to waste time on endless gossiping. When I was through swimming as hard as I could, all I wanted was to hurry home and read.

I went through the motions—shopping, cooking, playing with my son, having sex with my husband. It was easy once I got the hang of it. All I had to do was break the connection between my mind and my body. While my body went about its business, my mind floated in its own inner space. I ran the house without a thought in my head, feeding snacks to my son, chatting with my husband.

After I gave up sleeping, it occurred to me what a simple thing reality is, how easy it is to make it work. It's just reality. Just housework. Just a home. Like running a simple machine. Once you learn to run it, it's just a matter of repetition. You push this button and pull that lever. You adjust a gauge, put on the lid, set the timer. The same thing, over and over.

Of course, there were variations now and then. My mother-in-law had dinner with us. On Sunday, the three of us went to the zoo. My son had a terrible case of diarrhea.

But none of these events had any effect on my being. They swept past me like a silent breeze. I chatted with my mother-in-law, made dinner for four, took a picture in front of the bear cage, put a hot-water bottle on my son's stomach and gave him his medicine.

No one noticed that I had changed—that I had given up sleeping entirely, that I was spending all my time reading, that my mind was someplace a hundred years—and hundreds of miles—from reality. No matter how mechanically I worked, no matter how little love or emotion I invested in my handling of reality, my husband and my son and my mother-in-law went on relating to me as they always had. If anything, they seemed more at ease with me than before.

And so a week went by.

Once my constant wakefulness entered its second week, though, it started to worry me. It was simply not normal. People are supposed to sleep. All people sleep. Once, some years ago, I had read about a form of torture in which the victim is prevented from sleeping. Something the Nazis did, I think. They'd lock the person in a tiny room, fasten his eyelids open, and keep shining lights in his face and making loud noises without a break. Eventually, the person would go mad and die.

I couldn't recall how long the article said it took for the madness to set in, but it couldn't have been much more than three or four days. In my case, a whole week had gone by. This was simply too much. Still, my health was not suffering. Far from it. I had more energy than ever.

One day, after showering, I stood naked in front of the mirror. I was amazed to discover that my body appeared to be almost bursting with vitality. I studied every inch of myself, head to toe, but I could find not the slightest hint of excess flesh, not one wrinkle. I no longer had the body of a young girl, of course, but my skin had far more glow, far more tautness, than it had before. I took a pinch of flesh near my waist and found it almost hard, with a wonderful elasticity.

It dawned on me that I was prettier than I had realized. I looked so much younger than before that it was almost shocking. I could probably pass for twenty-four. My skin was smooth. My eyes were bright, lips moist. The shadowed area beneath my protruding cheekbones (the one feature I really hated about myself) was no longer noticeable—at all. I sat down and looked at my face in the mirror for a good thirty minutes. I studied it from all angles, objectively. No, I had not been mistaken: I was really pretty.

What was happening to me?

I thought about seeing a doctor.

I had a doctor who had been taking care of me since I was a child and to whom I felt close, but the more I thought about how he might react to my story, the less inclined I felt to tell it to him. Would he take me at my word? He'd probably think I was crazy if I said I hadn't slept in a week. Or he might dismiss it as a kind of neurotic insomnia. But if he did believe I was telling the truth, he might send me to some big research hospital for testing.

And *then* what would happen?

I'd be locked up and sent from one lab to another to be experimented on. They'd do EEGs and EKGs and urinalyses and blood tests and psychological screening and who knows what else.

I couldn't take that. I just wanted to stay by myself and quietly read my book. I wanted to have my hour of swimming every day. I wanted my freedom: That's what I wanted more than anything. I didn't want to go to any hospitals. And even if they *did* get me into a hospital, what would they find? They'd do a mountain of tests and formulate a mountain of hypotheses, and that would be the end of it. I didn't want to be locked up in a place like that.

One afternoon, I went to the library and read some books on sleep. The few books I could find didn't tell me much. In fact, they all had only one thing to say: that sleep is rest. Like turning off a car engine. If you keep a motor running constantly, sooner or later it will break down. A running engine must produce heat, and the accumulated heat fatigues the machinery itself. Which is why you have to let the engine rest. Cool down. Turning off the engine—that, finally, is what sleep is. In a human being, sleep provides rest for both the flesh and the spirit. When a person lies down and rests her muscles, she simultaneously closes her eyes and cuts off the thought process. And excess thoughts release an electrical discharge in the form of dreams.

One book did have a fascinating point to make. The author maintained that human beings, by their very nature, are incapable of escaping from certain fixed idiosyncratic tendencies, both in their thought processes and in their physical movements. People unconsciously fashion their own action- and thought-tendencies, which under normal circumstances never disappear. In other words, people live in the prison cells of their own tendencies. What modulates these tendencies and keeps them in check—so the organism doesn't wear down as the heel of a shoe does, at a particular angle, as the author puts it—is nothing other than sleep. Sleep therapeutically counteracts the tendencies. In sleep, people naturally relax muscles that have been consistently used in only one direction; sleep both calms and provides a discharge for thought circuits that have likewise been used in only one direction. This is how people are cooled down. Sleeping is an act that has been programmed, with karmic inevitability, into the human system, and no one can diverge from it. If a person *were* to diverge from it, the person's very "ground of being" would be threatened.

"Tendencies?" I asked myself.

The only "tendency" of mine that I could think of was housework—those chores I perform day after day like an unfeeling machine. Cooking and shopping and laundry and mothering: What were they if not "tendencies"? I could do them with my eyes closed. Push the buttons. Pull the levers. Pretty soon, reality just flows off and away. The same physical movements over and over. Tendencies. They were consuming me, wearing me down on one side like the heel of a shoe. I needed sleep every day to adjust them and cool me down.

Was that it?

I read the passage once more, with intense concentration. And I nodded. Yes, almost certainly, that *was* it.

So, then, what was this life of mine? I was being consumed by my tendencies and then sleeping to repair the damage. My life was nothing but a repetition of this cycle. It was going nowhere.

Sitting at the library table, I shook my head.

I'm through with sleep! So what if I go mad? So what if I lose my "ground of being"? I will not be consumed by my "tendencies." If sleep is nothing more than a periodic repairing of the parts of me that are being worn away, I don't want it anymore. I don't need it anymore. My flesh may have to be consumed, but my mind belongs to me. I'm keeping it for myself. I will not hand it over to anyone. I don't want to be "repaired." I will not sleep.

I left the library filled with a new determination.

Now my inability to sleep ceased to frighten me. What was there to be afraid of? Think of the advantages! Now the hours from ten at night to six in the morning belonged to me alone. Until now, a third of every day had been used up by sleep. But no more. No more. Now it was mine, just mine, nobody else's, all mine. I could use this time in any way I liked. No one would get in my way. No one would make demands on me. Yes, that was it. I had expanded my life. I had increased it by a third.

You are probably going to tell me that this is biologically abnormal. And you may be right. And maybe someday in the future I'll have to pay back the debt I'm building up by continuing to do this biologically abnormal thing. Maybe life will try to collect on the expanded part—this "advance" it is paying me now. This is a groundless hypothesis, but there is no ground for negating it, and it feels right to me somehow. Which means that in the end, the balance sheet of borrowed time will even out.

Honestly, though, I didn't give a damn, even if I had to die young. The best thing to do with a hypothesis is to let it run any course it pleases. Now, at least, I was expanding my life, and it was wonderful. My hands weren't empty anymore. Here I was—alive, and I could feel it. It was real. I wasn't being consumed any longer. Or at least there was a part of me in existence that was not being consumed, and that was what gave me this intensely real feeling of being alive. A life without that feeling might go on forever, but it would have no meaning at all. I saw that with absolute clarity now.

After checking to see that my husband was asleep, I would go sit on the living-room sofa, drink brandy by myself, and open my book. I read *Anna Karenina* three times. Each time, I made new discoveries. This enormous novel was full of revelations and riddles. Like a Chinese box, the world of the novel contained smaller worlds, and inside those were yet smaller worlds. Together, these worlds made up a single universe, and the universe waited there in the book to be discovered by the reader. The old me had been able to understand only the tiniest fragment of it, but the gaze of this new me could penetrate to the core with perfect understanding. I knew exactly what the great Tolstoy wanted to say, what he wanted the reader to get from his book; I could see how his message had organically crystallized as a novel, and what in that novel had surpassed the author himself.

No matter how hard I concentrated, I never tired. After reading *Anna Karenina* as many times as I could, I read Dostoyevski. I could read book after book with utter concentration and never tire. I could understand the most difficult passages without effort. And I responded with deep emotion.

I felt that I had always been meant to be like this. By abandoning sleep I had expanded myself. The power to concentrate was the most important thing. Living without this power would be like opening one's eyes without seeing anything.

Eventually, my bottle of brandy ran out. I had drunk almost all of it by myself. I went to the gourmet department of a big store for another bottle of Rémy-Martin. As long as I was there, I figured, I might as well buy a bottle of red wine, too. And a fine crystal brandy glass. And chocolate and cookies.

Sometimes while reading I would become overexcited. When that happened, I would put my book down and exercise—do calisthenics or just walk around the room. Depending on my mood, I might go out for a nighttime drive. I'd change clothes, get into my Civic, and drive aimlessly around the neighborhood. Sometimes I'd drop into an all-night fast-food place for a cup of coffee, but it was such a bother to have to deal with other people that I'd usually stay in the car. I'd stop in some safe-looking spot and just let my mind wander. Or I'd go all the way to the harbor and watch the boats.

One time, though, I was questioned by a policeman. It was two-thirty in the morning, and I was parked under a streetlamp near the pier, listening to the car stereo and watching the lights of the ships passing by. He knocked on my window. I lowered the glass. He was young and handsome, and very polite. I explained to him that I couldn't sleep. He asked for my license and studied it for a while. "There was a murder here last month," he said. "Three young men attacked a couple. They killed the man and raped the woman." I remembered having read about the incident. I nodded. "If you don't have any business here, ma'am, you'd better not hang around here at night." I thanked him and said I would leave. He gave me my license back. I drove away.

That was the only time anyone talked to me. Usually, I would drift through the streets at night for an hour or more and no one would bother me. Then I would park in our underground garage. Right next to my husband's white Sentra; he was upstairs sleeping soundly in the darkness. I'd listen to the crackle of the hot engine cooling down, and when the sound died I'd go upstairs.

The first thing I would do when I got inside was check to make sure my husband was asleep. And he always was. Then I'd check my son, who was always sound asleep, too. They didn't know a thing. They believed that the world was as it had always been, unchanging. But they were wrong. It was changing in ways they could never guess. Changing a lot. Changing fast. It would never be the same again.

One time, I stood and stared at my sleeping husband's face. I had heard a thump in the bedroom and rushed in. The alarm clock was on the floor. He had probably knocked it down in his sleep. But he was sleeping as soundly as ever, completely unaware of what he had done. What would it take to wake this man? I picked up the clock and put it back on the night table. Then I folded my arms and stared at my husband. How long had it been—years?—since the last time I had studied his face as he slept?

I had done it a lot when we were first married. That was all it took to relax me and put me in a peaceful mood. I'll be safe as long as he goes on sleeping peacefully like this, I'd tell myself. Which is why I spent a lot of time watching him in his sleep.

But, somewhere along the way, I had given up the habit. When had that been? I tried to remember. It had probably happened back when my mother-in-law and I were sort of quarreling over what name to give my son. She was big on some religious cult kind of thing, and had asked her priest to "bestow" a name on the baby. I don't remember exactly the name she was given, but I had no intention of letting some priest "bestow" a name on my child. We had some pretty violent arguments at the time, but my husband couldn't say a thing to either of us. He stood by and tried to calm us.

After that, I lost the feeling that my husband was my protector. The one thing I thought I wanted from him he had failed to give me. All he had managed to do was make me furious. This happened a long time ago, of course. My mother-in-law and I have long since made up. I gave my son the name I wanted to give him. My husband and I made up right away, too.

I'm pretty sure that was the end, though, of my watching him in his sleep.

So there I stood, looking at him sleeping as soundly as always. One bare foot stuck out from under the covers at a strange angle— so strange that the foot could have belonged to someone else. It was a big, chunky foot. My husband's mouth hung open, the lower lip drooping. Every once in a while, his nostrils would twitch. There was a mole under his eye that bothered me. It was so big and vulgar-looking. There was something vulgar about the way his eyes were closed, the lids slack, covers made of faded human flesh. He looked like an absolute fool. This was what they mean by "dead to the world." How incredibly ugly! He sleeps with such an ugly face! It's just too gruesome, I thought. He couldn't have been like this in the old days. I'm sure he must have had a better face when we were first married, one that was taut and alert. Even sound asleep, he couldn't have been such a blob.

I tried to remember what his sleeping face had looked like back then, but I couldn't do it, though I tried hard enough. All I could be sure of was that he *couldn't* have had such a terrible face. Or was I just deceiving myself? Maybe he had always looked like this in his sleep and I had been indulging in some kind of emotional projection. I'm sure that's what my mother would say. That sort of thinking was a specialty of hers. "All that lovey-dovey stuff lasts two years—three years tops," she always used to insist. "You were a new bride," I'm sure she would tell me now. "Of *course*, your little hubby looked like a darling in his sleep."

I'm sure she would say something like that, but I'm just as sure she'd be wrong. He *had* grown ugly over the years. The firmness had gone out of his face. That's what growing old is all about. He was old now, and tired. Worn out. He'd get even uglier in the years ahead, that much was certain. And I had no choice but to go along with it, put up with it, resign myself to it.

I let out a sigh as I stood there watching him. It was a deep sigh, a noisy one as sighs go, but of course he didn't move a muscle. The loudest sigh in the world would never wake him up.

I left the bedroom and went back to the living room. I poured myself a brandy and started reading. But something wouldn't let me concentrate. I put the book down and went to my son's room. Opening the door, I stared at his face in the light spilling in from the hallway. He was sleeping just as soundly as my husband was. As he always did. I watched him in his sleep, looked at his smooth, nearly featureless face. It was very different from my husband's: It was still a child's face, after all. The skin still glowed; it still had nothing vulgar about it.

And yet, something about my son's face annoyed me. I had never felt anything like this about him before. What could be making me feel this way? I stood there, looking, with my arms folded. Yes, of course I loved my son, loved him tremendously. But still, undeniably, that something was bothering me, getting on my nerves.

I shook my head.

I closed my eyes and kept them shut. Then I opened them and looked at my son's face again. And then it hit me. What bothered me about my son's sleeping face was that it looked exactly like my husband's. And exactly like my mother-in-law's. Stubborn. Self-satisfied. It was in their blood—a kind of arrogance I hated in my husband's family. True, my husband is good to me. He's sweet and gentle and he's careful to take my feelings into account. He's never fooled around with other women, and he works hard. He's serious, and he's kind to everybody. My friends all tell me how lucky I am to have him. And I can't fault him, either. Which is exactly what galls me sometimes. His very absence of faults makes for a strange rigidity that excludes imagination. That's what grates on me so.

And that was exactly the kind of expression my son had on his face as he slept.

I shook my head again. This little boy is a stranger to me, finally. Even after he grows up, he'll never be able to understand me, just as my husband can hardly understand what I feel now.

I love my son, no question. But I sensed that someday I would no longer be able to love this boy with the same intensity. Not a very maternal thought. Most mothers never have thoughts like that. But as I stood there looking at him asleep, I knew with absolute certainty that one day I would come to despise him.

The thought made me terribly sad. I closed his door and turned out the hall light. I went to the living-room sofa, sat down, and opened my book. After reading a few pages, I closed it again. I looked at the clock. A little before three.

I wondered how many days it had been since I stopped sleeping. The sleeplessness started the Tuesday before last. Which made this the seventeenth day. Not one wink of sleep in seventeen days.

Seventeen days and seventeen nights. A long, long time. I couldn't even recall what sleep was like.

I closed my eyes and tried to recall the sensation of sleeping, but all that existed for me inside was a wakeful darkness. A wakeful darkness: What it called to mind was death.

Was I about to die?

And if I died now, what would my life have amounted to?

There was no way I could answer that.

All right, then, what *was* death?

Until now, I had conceived of sleep as a kind of model for death. I had imagined death as an extension of sleep. A far deeper sleep than ordinary sleep. A sleep devoid of all consciousness. Eternal rest. A total blackout.

But now I wondered if I had been wrong. Perhaps death was a state entirely unlike sleep, something that belonged to a different category altogether—like the deep, endless, wakeful darkness I was seeing now.

No, that would be too terrible. If the state of death was not to be a rest for us, then what was going to redeem this imperfect life of ours, so fraught with exhaustion? Finally, though, no one knows what death is. Who has ever truly seen it? No one. Except the ones who are dead. No one living knows what death is like. They can only guess. And the best guess is still a guess. Maybe death *is* a kind of rest, but reasoning can't tell us that. The only way to find out what death is is to die. *Death can be anything at all.*

An intense terror overwhelmed me at the thought. A stiffening chill ran down my spine. My eyes were still shut tight. I had lost the power to open them. I stared at the thick darkness that stood planted in front of me, a darkness as deep and hopeless as the universe itself. I was all alone. My mind was in deep concentration, and expanding. If I had wanted to, I could have seen into the uttermost depths of the universe. But I decided not to look. It was too soon for that.

If death was like this, if to die meant being eternally awake and staring into the darkness like this, what should I do?

At last, I managed to open my eyes. I gulped down the brandy that was left in my glass.

I'm taking off my pajamas and putting on jeans, a T-shirt, and a windbreaker. I tie my hair back in a tight ponytail, tuck it under the windbreaker, and put on a baseball cap of my husband's. In the mirror, I look like a boy. Good. I put on sneakers and go down to the garage.

I slip in behind the steering wheel, turn the key, and listen to the engine hum. It sounds normal. Hands on the wheel, I take a few deep breaths. Then I shift into gear and drive out of the building. The car is running better than usual. It seems to be gliding across a sheet of ice. I ease it into higher gear, move out of the neighborhood, and enter the highway to Yokohama.

It's only three in the morning, but the number of cars on the road is by no means small. Huge semis roll past, shaking the ground as they head east. Those guys don't sleep at night. They sleep in the daytime and work at night for greater efficiency.

What a waste. I could work day *and* night. I don't have to sleep.

This is biologically unnatural, I suppose, but who really knows what is natural? They just infer it inductively. I'm beyond that. A priori. An evolutionary leap. A woman who never sleeps. An expansion of consciousness.

I have to smile. A priori. An evolutionary leap.

Listening to the car radio, I drive to the harbor. I want classical music, but I can't find a station that broadcasts it at night. Stupid Japanese rock music. Love songs sweet enough to rot your teeth. I give up searching and listen to those. They make me feel I'm in a far-off place, far away from Mozart and Haydn.

I pull into one of the white-outlined spaces in the big parking lot at the waterfront park and cut my engine. This is the brightest area of the lot, under a lamp, and wide open all around. Only one car is parked here—an old white two-door coupe of the kind that young people like to drive. Probably a couple in there now, making love—no money for a hotel room. To avoid trouble, I pull my hat low, trying not to look like a woman. I check to see that my doors are locked.

Half-consciously, I let my eyes wander through the surrounding darkness, when all of a sudden I remember a drive I took with my boyfriend the year I was a college freshman. We parked and got into some heavy petting. He couldn't stop, he said, and he begged me to let him put it in. But I refused. Hands on the steering wheel, listening to the music, I try to bring back the scene, but I can't recall his face. It seems to have happened such an incredibly long time ago.

All the memories I have from the time before I stopped sleeping seem to be moving away with accelerating speed. It feels so strange, as if the me who used to go to sleep every night is not the real me, and the memories from back then are not really mine. This is how people change. But nobody realizes it. Nobody notices.

Only *I* know what happens. I could try to tell them, but they wouldn't understand. They wouldn't believe me. Or if they did believe me, they would have absolutely no idea what I'm feeling. They would only see me as a threat to their inductive worldview.

I am changing, though. *Really* changing.

How long have I been sitting here? Hands on the wheel. Eyes closed. Staring into the sleepless darkness.

Suddenly I'm aware of a human presence, and I come to myself again. There's somebody out there. I open my eyes and look around. Someone is outside the car. Trying to open the door. But the doors are locked. Dark shadows on either side of the car, one at each door. Can't see their faces. Can't make out their clothing. Just two dark shadows, standing there.

Sandwiched between them, my Civic feels tiny—like a little pastry box. It's being rocked from side to side. A fist is pounding on the right-hand window. I know it's not a policeman. A policeman would never pound on the glass like this and would never shake my car. I hold my breath. What should I do? I can't think straight. My underarms are soaked. I've got to get out of here. The key. Turn the key. I reach out for it and turn it to the right. The starter grinds.

The engine doesn't catch. My hand is shaking. I close my eyes and turn the key again. No good. A sound like fingernails clawing a giant wall. The motor turns and turns. The men—the dark shadows—keep shaking my car. The swings get bigger and bigger. They're going to tip me over!

There's something wrong. Just calm down and think, then everything will be okay. Think. Just think. Slowly. Carefully. Something is wrong.

Something is wrong.

But what? I can't tell. My mind is crammed full of thick darkness. It's not taking me anywhere. My hands are shaking. I try pulling out the key and putting it back in again. But my shaking hand can't find the hole. I try again and drop the key. I curl over and try to pick it up. But I can't get hold of it. The car is rocking back and forth. My forehead slams against the steering wheel.

I'll never get the key. I fall back against the seat, cover my face with my hands. I'm crying. All I can do is cry. The tears keep pouring out. Locked inside this little box, I can't go anywhere. It's the middle of the night. The men keep rocking the car back and forth. They're going to turn it over.

—translated by Jay Rubin

Anthony Schneider

You Sleep

on the Hilltop

Killing, parties, insomnia. For me they happened all at once.

It was the year of the massacres in Soweto and Queenstown and Uitenhage. Police opened fire outside churches, drove through funeral marches in armored vehicles, and hid in unmarked vans, shooting at women and children as they tried to flee. It was the year of grenades, *sjamboks*, rifles, work stoppages, riots, and more funerals than anyone could count. It was also the year my parents started throwing parties.

It was an unconscious reaction, like nervous laughter. Everyone drank too much, and the guests spilled outside into the courtyard, where they sipped gin and tonics under the swaying bibrikos vines and talked about rugby, who was performing next at Sun City, and whether to holiday in Clifton or Plettenberg Bay.

I lay awake in bed at night, terrified by shadows and dreams. There was a heavy stinkwood chest in my room, and even when its doors were closed, horrible phantoms could slither out and slide across the floor and sink a thousand teeth into me. You, Gogo, my nanny, were the only one who could keep night's demons at bay. Your Zulu lullabies made night safe. You sang for me in your soft, resiny, patient voice—the sound of sleep. I would close my eyes and listen, until the song you sang was all I heard, all I thought, everything. *We ngane ulele eziweni,* you sang. You sleep on the hilltop. And I slept.

The parties were always crowded. By the time everyone had arrived, the living room would be humming with a polyphony of sounds— conversations, ice clinking in glasses, bursts of laughter, and the jazz records Dad put on early in the evening. Mom would sit with an eager group on the leather couches in the middle of the room; Dad would hover near the bar, patting backs, pouring drinks, offering cigars from his humidor.

You didn't like the parties. You would trail Gideon with a tray of clean glasses, alert and uneasy in your white apron. You waited impatiently for him to remove the glasses, then tucked the tray under your arm and hurried out through the courtyard to the back door of the kitchen.

Sometimes you would try to draw me away. "Read me a story while I am cooking," you would say. Or "Show me your homework."

But you couldn't devote yourself to me, and there was no one else for me to play with, because Mom and Dad didn't allow me to invite a friend to sleep over when they threw a party. So I usually ended up in the living room, which, even though it was full of drunk, sloppy grown-ups, held more appeal than solitude did.

They threw a party the night before Easter Sunday. It was the school holidays, and I was allowed to stay up later than usual. I wandered about the living room, chest-high among the revelers. Gideon wove through the crowd in his starched waiter's uniform, offering plates of salmon and cucumber sandwiches. The din of conversation and laughter grew louder and louder until it subsumed Dad's John Coltrane record. Clouds of blue smoke fanned out across the ceiling, and the smell of cigars and perfume mingled with the dry, grassy breeze coming in through the window.

I was standing alone at the window when Carlo d'Amico, the Italian painter, came up behind me, his big, rough face glowing under his white hair and beard.

"*Andiamo*," he said, his breath thick with red wine. "Let's go and look at the stars." We walked through the courtyard beneath the vines that dangled like fingers from the wooden beams, across the dark grass between the pool and *koppie* until the party sounded like a radio playing far away. Dry grass cracked under my shoes; the spilled, luminous moon hovered behind the treetops. Carlo breathed in deeply, tilting his head back and throwing his hands in the air.

"*Que bello il cielo*," he said, and stared knowingly into the stars, as if he recognized an old friend standing up there. He raised his arms and turned around slowly. Then he looked at me and grinned. *"Fa peepee?"*

I nodded, and we tiptoed around the bougainvilleas and un-buttoned our flies and pissed into the moonlit night, hidden by trees so dark they looked like they had been cut out of the sky.

Back inside, Mom told me to go into the kitchen and ask if the roast beef was ready. "People are getting hungry," she said. Then a frown crossed her face. "Did you eat?"

"Of course."

When I came into the kitchen, you were on the phone. You stood at the back door, your chest rising and falling slowly against the tangled knots of aloes that covered the *koppie,* your *doek*-cov-ered head framed by cloudy sky. Your fingers moved across your waist, folding and refolding the dishcloth you kept tucked into your apron.

"Gogo."

My voice startled you. "What is it, little one?"

"Mom wants to know if the roast beef is ready."

Your eyes were focused elsewhere, far away, and for an instant you said nothing. Then you spoke quickly into the phone and hung up.

"What's wrong?" I asked.

You looked at the phone, then back at me. "It's Simon. He's sick. That was Janie."

"What kind of sick?"

"She doesn't know," you said. "She only just found out herself."

I knew so little about Simon or your other children. He was living with your sister, Jane, attending high school. He was a good student, and you wanted him to finish school, which he would not have been able to do in Natal, where he lived with your brother's family. So he went to Soweto, even though you didn't like the townships, to get an education. You were proud of him. He made good marks at school. You said he was going to be a teacher one day. "He's clever like you," you often told me. "Both my boys are so clever." And you would smile, the corners of your mouth squeezing bulges in your plump, soft cheeks.

I first met Simon two years earlier. I didn't expect the bleak, anonymous look of poverty I saw in his face and clothes—the threadbare sweater and big secondhand shoes that made him walk in long, exaggerated steps. His hair was cut so short that bits of scalp were visible through the black millimeter of stubble. His wide eyes reminded me of yours, Gogo, and he had the same full, pursed lips that made him look as if he were always about to speak, even though he was so reticent. We circled each other warily at first. He was older, but I had every other advantage. After a while, though, we forged a temporary friendship. You gave us a chore to do— cutting the ends off green beans—and when we were finished I asked if he wanted to ride my bicycle. After that I showed him my model planes and we played soccer together, dribbling the ball around the front lawn for over an hour. When I looked up at the house I saw you standing in the living room, watching through the window.

Now you wiped the wooden platter piled high with sliced beef, took a glass jar of mustard out of the fridge, and carried the platter out of the kitchen, backing through the swinging door. I followed.

Dad was at his habitual perch, red-faced and loud, leaning on one hand, thrusting cigars at people with the other. Always the salesman. When he saw me looking, he smiled sheepishly and shrugged.

"How's my big boy?"
"Fine."

When I looked up you were already heading back to the kitchen. I sat near the door to the courtyard and listened to a conversation about American cities I'd never heard of—Atlanta, Houston, Miami. People drifted around me. The tray of beef was surrounded by a hungry throng. The side tables were littered with empty glasses and ashtrays piled with lipstick-ringed cigarette butts and stumps of cigars. At the bar, empty bottles were stacked behind the pewter ice bucket; two flies circled sluggishly above the plate of cut lemons.

The guests shifted gears. Ties were loosened, jackets flung on the backs of chairs, high heels kicked off, and now I heard hiccups and slurred speech all around me. Our neighbor, Mr. Murphy, laughed too loudly, slapping his knee, and then stopped suddenly and looked around bewildered, as if he had been plopped into an unfamiliar place and didn't recognize his surroundings.

There was a joke that year about how the actions of South Africa's politicians were tantamount to rearranging the deck chairs on the *Titanic*. It was something I heard often, a familiar liberal riposte, and I was proud of myself for understanding and laughed every time I heard it, to show off. And that night when the party was at its peak, I imagined our house as a ship, the dark slopes of aloes behind us crests of waves, the front lawn an endless sea.

. . .

"Say good night to medem and master." You conducted me to my parents, who each gave me a peck on the cheek, and then we made our way out of the living room. I followed your hefty calves wrapped in thick brown socks up the stairs, until the sounds of the party were faint and the only smell was you, the menthol of your snuff, the faint scent of cheap soap, your warm buttery skin.

You sat on the desk chair and watched me get ready for bed. Above you, my posters filled the wall—a fake bullfight announcement with my name in big black letters, Team Elf, the Concorde in an empty sky. The windows glinted in a slant of moonlight, and shadows sprang from the bars, crisscrossing your face and covering the floor in long black stripes. You were distracted that night—somber, ponderous. My desk was littered with parts of model airplanes, glue, and scraps of newspaper, and you moved things about absently while you waited, making piles.

"Do you drink, Gogo?"

"*Haikona.* No. Of course not." Your face became stern, your forehead creasing with frowns. "When I was younger. But not anymore. Come, time to go to sleep now."

Obediently I lay down and pulled the covers up to my chin. The wind whistled through dry branches; thorn trees scraped against drainpipes. I wanted to feel sleepy, but terrors lurked. I saw hands holding daggers rising from the shadows, sliding through the curtains.

"Just one song tonight."

"Why? They'll be okay downstairs without you. Everyone's drunk."

"But I must finish cleaning. I have to leave early tomorrow morning."

"Why?"

"I must go and see Simon. Lie down now."

You straightened the duvet around me, then picked up your baggy apron, and as you sat down on the edge of the bed, kitchen smells—butter and roast beef—floated out. In the dim room, I saw only the outline of your face as you gathered the lullaby inside you, then started to hum, quietly at first, your voice as low as an oboe. Gradually, your humming grew louder and louder, your lips pressed tightly together. And then you paused, took a deep breath, and started to sing the words. As usual, I drifted off, even as I tried to listen to the song. Opening my heavy-lidded eyes, I saw you beside me, your lips like cupped hands, singing.

I heard the rustle of your apron as you tiptoed out of the room.

I slept fitfully that night. Waves of laughter from downstairs entered my dreams as howling screams, and every time I heard footsteps outside I was sure they came from the roof, and I lay stiffly in the darkness, waiting for a face to appear in the moonlit window. The familiar ripple of shadows across the wall became a sinister flock of birds, all beaks and claws.

Nighttime was awful without you. A shirt draped across a chair could grow arms and sprout a maniac head. Piles of books on the desk became squat men with guns in their hands.

You were away nearly a week, and the nights were cold and long. My parents went out just about every night, and without you there the hours were empty. I watched TV, played with my remote-controlled car, tossed pinecones into the fire. Gideon would sit in the kitchen, drinking milky tea from his enamel mug, babysitting from a distance.

I asked Mom what was wrong with Simon. She said he fell off his bicycle. "Don't worry. He'll be fine. And Gogo will be back soon."

I imagined you sitting beside him, dabbing cuts and bruises with Dettol, telling him stories, singing him lullabies.

One night I asked Mom to sing me a lullaby, and she sang "Embraceable You," but her voice was reedy and too sharp to make me sleepy, and she left the room before I had even closed my eyes. She said I shouldn't be afraid. She said nighttime was just the same as daytime, only dark. I said I wasn't afraid; I just liked lullabies.

Death was all around that year. There were riots, murders, funerals. We didn't talk about them, but killings were reported almost daily on TV. That winter thirty thousand people attended the funeral of Matthew Goniwe and three United Democratic Front members in the tiny town of Cradock, not far from where your daughter Mary lived, and P. W. Botha appeared on television to announce a State of Emergency. Thousands were beaten, tortured, wounded, killed. I asked my parents why it was happening, but they were evasive. They didn't want to know, didn't want to imagine. Better to get drunk and forget.

But I wasn't drunk; I was terrified. How could we be safe? Wouldn't the angry masses who saw death close up, who had lost fathers and mothers, sisters and brothers, finally rise up? Couldn't the neighbor's servant, the milkman, the grocery delivery boy bring the violence into the white suburbs and kill just anybody? Even the crazed white "security forces" who shot township kids in the back— couldn't they kill me? I imagined the eye of a rifle aimed at me, and wondered what it felt like to die. My heart banged with jagged terror as I contemplated the endless blackness that came after life ended.

I hardly slept. Wheezing winds and howling dogs kept me awake, and my sleep was choked by nightmares. I dreamed of willowy black men with sharp *tsotsi*'s knives in their teeth climbing through the bars of my window and of explosions and rocks raining down through the jacaranda trees. I would be trapped in a house on fire, running but not moving, as flames crept up the walls and plaster dropped from the ceiling and thumped onto the floorboards in puffy white clouds. I dreamed of the quick, glinting flash of blades, and of a river of blood.

I would jerk awake and look for you, Gogo, hoping to see your face, your big, black, calming eyes, and hear your soft songs. But then I would realize you were gone, and I was alone in the bleak, black night. I wanted you back; I didn't care about Simon.

The last time he had visited us, he stayed only a few hours. He was on his way home to Natal for the Christmas holidays. He had grown, aged, and his face was stern—a man's face. Even though it was hot, he wore gray flannel pants, probably once a suit of Dad's given on Boxing Day or rescued from the garbage. He waited in the servant's quarters instead of coming into the kitchen, and when I asked if he wanted to go swimming, he said no. I had to ask several times before he agreed to go into the *koppie*, where I showed him my secret stash of marbles and firecrackers. I made a clearing in the brush and lit a Roman candle. He jumped back when it caught, spraying a rainbow of light, but then he clapped his hands in delight, and we watched it spin and flare, a flower of bright, burning color in the dry scrub.

. . .

You came back at night. It was raining, a fierce rain that banged like a thousand hammers on the roof and drainpipes and rattled doors and shutters. I didn't hear you as you crept into my room, but when I opened my eyes, you were there beside me.

"Gogo. You're back."

"I'm back." You sat on the edge of the bed, stroked my hair, and took my hand in yours. "We'll stay together now and keep scary things away."

Thunder boomed and lightning flashed in the sky, covering the room with jagged streaks. The lightning lit your face briefly, and I saw there were tears on your cheeks.

"Gogo?"

"*Hai, hai, hai.*" You shook your head slowly and clamped a hand over your mouth.

After a while you sat up and patted your cheeks dry, and made sure your *doek* was straight. When you rubbed your hands together, it was as if you were feeling your own skin for the first time. You fingered the sleeves of your woolen jersey and pursed your lips tightly. Rain swooshed down the windows, casting shadow trails on the posters and walls.

When you spoke at last it was in an unfamiliar tone—soft, timorous, hesitant. "I'm sad, little one. You know why I went home?"

"Simon—"

"Simon is dead. My boy was killed."

"What?"

I found out later that there was a fight in the *shebeen* down the road from Janie's house, and he happened to be riding past on his bicycle when the police began shooting.

That night you sang and sang, until the lightning and thunder stopped and I couldn't tell the song from the sound of rain and wind.

We ngane ulele eziweni.

Ameena Meer

M A N N E Q U I N

Jamila said "Yes" after three days. Three mornings being woken up by the flutters in her stomach. Three afternoons where the flush seemed to plump her cheeks and lips hot and red with liquid. Three exhausting nights flopping in her bed, dreaming of the café au lait wall-to-wall carpeting and upholstered Italian chairs in tones of hazelnut and almond. Three mornings contemplating champagne Mercedeses and leather seats and automatic CD changers while her breakfast cereal lost its crunch and turned the milk pink. For those three days, she left the table without swallowing more than a few spoonfuls of anything. After seventy-two hours of living on fantasy (and no sleep and very little nutrition), she felt as if she were walking on bubbles, like she was made out of air and glass. When she tried to read the morning paper, her brain kept floating away like a helium balloon, her thoughts bobbing in and out of the fluffy clouds.

Even the sharp ring of the telephone, the pointed bleep of the answering machine, the confident swagger in his voice (being recorded), didn't pierce her mind or deflate her images of herself—with a suntan, wearing pale lipstick and heavy gold earrings, a beige cardigan over a gym-trained torso in linen slacks—the sleek suburban wife. It was only her father's voice, reminding her to call him back, that disturbed her revery.

Her parents said, "We're-so-happy-for-you-both," that first evening. He had come over with Christmas presents: an orchid plant, a designer teapot, a brown paper bag of gourmet tea, a brushed-steel vase. All objects denoting taste, wealth, a certain contemporary aesthetic sensibility, and a sensitivity to her parents' own style of living.

Jamila was flattered, but angry at her parents for giving in so easily. She didn't think she was such a bad deal. She thought her father should have dragged him across the coals (just a little), made him squirm on the sofa. She wanted to see that smug grin be replaced by an uncomfortable smile—have him question, just for a second, what he seemed so sure was already his—she wanted him off balance. So they could go back to the even score of the night she met him. And why didn't her mother first make inquiries on the Indo-Pakistani telephone circuit? Confirming his true financial status and social standing, not to mention his previous love life. Or dig out any secrets. They did know that he'd been divorced, his previous marriage was to an English girl. No one took that too seriously anyway and, in the long run, it was considered to be an asset in an Indian man. Everyone knew how demanding those modern European girls were. It meant that he'd been broken in already, so to speak.

In fact, by the time they got home from the two-hour car ride, where her mother sat in the passenger seat of the Mercedes beside her future fiancé and discussed Jamila's future (while Jamila sat silently in the back seat, watching the Christmas lights on the houses making strange geometric shapes in the darkness), the deal had been made.

Jamila burst into tears as she brushed her teeth.

Her mother came into the bathroom and said, "Oh, poor thing, what's wrong?"

And Jamila said, "I can't believe that you just agreed, just like that, on the spur of the moment. You don't know anything about him, do you?" Her mother tried to defend herself by saying that she'd asked Jamila, in the car, whether or not she wanted to marry him. "You wanted me to tell him, right there? I've only met the guy three times. You were supposed to say, 'We'll talk about it . . .'" Jamila collapsed in a sobbing heap against the bathroom scale.

"I'm so sorry, Jamila," apologized her mother. "I really didn't know what to do. Shall I call him now and say, 'We've got to wait'?"

"No," sulked Jamila, turning her head and spitting a mouthful of toothpaste into the bathtub. "That'll just sound stupid." She stood up and started to wash her face with careful determination, pretending that the humiliation was too great for her to utter another word.

Her mother said, "But, you know, we already know all about him. I've heard about him from Bashan auntie and Nasreen and Alia. They all say he's very good. He's doing very well in his job and he plays tennis with Liaquat Iqbal, you know, the head of the South Asian division." But her mother did make the phone call and quite a few others and every one added to the glowing reports. Therefore, three days after being asked for her hand, Jamila said "Yes," and it was all set up. Her mother left for India the next week. She had to get Jamila's clothes made and to meet his parents.

That was how, by the end of the Christmas holidays, when she was back in New York, back at her job and ensconced in her tiny apartment—with only three days' contemplation—Jamila was engaged. She had a 22-karat gold ring set with five quarter-carat diamonds, modeled on her mother's own engagement ring. It was about then, or a week before, when her nightmares started.

Jamila was no longer kept awake by images of her new life. Rather, the excitement of the holidays and then the rush back at the office after the break had drained her. She was so tired by the end of the day that she usually fell asleep on the second exhale, her open mouth making a wet spot of saliva on her feather pillow. It was then, or what seemed like minutes later, that the visions started. From the night they began, Jamila found herself trapped in her dreams, unable to resolve them or to escape into consciousness, no matter how frightening the images.

When she was a child, Jamila's mother or father would wake her up when her nightmares made her scream in her sleep. A few minutes in the lucidity of light and a parental kiss would return her to a happier REM state. Since her engagement, however, her visions could no longer be exorcised. One night, Jamila opened her eyes and still found her body clawed and bleeding, the warm stain spreading scarlet on the sheet, the torn flesh on her face still sting-ing, and the howl of her attacker, though at bay, still clear and penetrating. It took more than an hour of lying semiconscious in bed for the hallucination to fade. Another morning, she woke up unable to lift her head or move her limbs, paralyzed, like a rabbit in the headlights of an oncoming car, while the sound of approaching footsteps echoed through the floor. Sometimes, in the dawn haze, she could still see the monster, dark and misshapen, emerging out of the wall toward her.

Jamila had been to university and, like everyone else post-psychoanalysis, could solve the simple puzzle of her nightmares. Who else was her pursuer but her future husband? All the monsters were only symbols. She must feel trapped by her decision. And, she admitted to herself, even awake, she found the man predatory. She kept being reminded of the first night they had dinner together, when she was gaining the advantage, when she danced around the granite kitchen floor to avoid his hands and eyes. They'd kissed. He'd attempted a seduction. She stopped it by saying she was wait-ing until she was married. She admitted that that had never been a reason for anything before. But now that she'd agreed to having an arranged marriage, she decided she'd better go the whole hog. He looked incredulous. "If your parents find someone for you to marry, you'll do it?" She nodded and pulled down her shirt. He grinned.

Still, she couldn't think of any tangible reason, with her eyes open, why she shouldn't marry him. She tried broaching the subject with her father on the phone. "I don't know," she said, in one of their Sunday evening discount long-distance phone calls, "I can't think of what to say to him. He keeps saying we have to make plans and I don't know what to say."

"Just be natural," advised her father. He was a doctor and tended to think rationally. "It's normal to feel a little nervous when

you're thinking about spending the rest of your life with someone."
Thinking about it made Jamila feel nauseated. Nervous like a small
animal blinded by the underbrush, with the barrel of an unseen rifle
aimed at its head and the smell of gunpowder in the air.

Jamila met Pierre at a chic gallery AIDS benefit. She was actually
meant to be checking invitations or making sure people gave dona-
tions, and he hadn't concerned himself with either. When she murmured, discreetly, so as not to upset the salon-style setting (someone was reading poetry), whether he had bought a ticket or would
like to make a larger donation, cash or checks were all right, he just
grinned. Then he unzipped his jacket. It was a multicolored nylon-
and-fiberfill monstrosity that blazed cheapness and lack of taste in
the midst of all the matte black and muted tones of the gallery. He
dropped the jacket along with an equally ugly nylon rucksack on
the polished wood floor. He leaned against the wall, dangerously
close to a painting, his head tilted back as if listening to the reader.
His hair didn't look especially clean and Jamila hoped he wouldn't
leave a greasy halo behind him. When she gestured again, his grin
reappeared. It was a wide, white-toothed, freckled-face grin like that
of a kid in a cornflakes commercial.

He didn't strike her as handsome. Or not so beautiful that she
was willing to forgive his ticket. But his smile was so earnest and
open that she liked him immediately. There was something fresh
about him in the gray city, something that made him seem clean
and honest. Trustworthy enough for her to give her phone number
to a complete stranger. He even walked her home afterward—

rather, she even let him walk her home afterward. He trotted along beside her down the sidewalk, winding around the piles of packing boxes and rubbish and artists' and dressmakers' detritus.

Jamila knew better than to do something like that. Urban girls learn quickly—or suffer the results—about the risks they take. She knew it was unsafe. The warning rang in her brain the same way it did on those nights when she left that oblong ketchup package unopened in the breast pocket of her denim jacket, and it kept calling her, drowning out her own moans. The innocuous-looking ones turn out to be the serial killers. But, as on those other life-threatening encounters, she let him walk her all the way to the door—now he knew where she lived! She let him kiss her good night, on both cheeks, the French way, so that she could inhale the intoxicating smell of male skin to savor later. Then she gently closed the door behind her and went to bed. Alone. She was not yet so stupid as to invite him up for a "tisane." But, buried under the sheets, she pulled out the smell and the unshaven scratch of his face, and imagined the feel of his body in bed beside her—the hollow spot between breast bones, the dark hairs that run like a path leading from the belly button down to the pubic hair—as she fell back into her nightmares.

Of course, Pierre telephoned the next evening as she walked in the door from work. *"Allo?"*—gentle and hesitant—and she said, "Hi, how are you?" right away. Saving him the embarrassment of having to identify himself. As a rule, she didn't do that for those first calls. She knew, by instinct again, that it was always better to let them get on the wrong foot from the beginning. It meant that you had the edge. But she was feeling unbalanced these days and decided to dispense with the games. In any case, she had nothing to lose. Nothing to win either. She was no longer in the game. She was twenty-two years old and engaged. No longer a crazy American girl. She was now the good Indian wife. The appeal of Pierre's French accent was just nostalgia, a longing for the affair she had last summer in Paris. She wanted another taste of that kind of overwhelming passion, something so intense she could barely breathe. So overwhelming that it was impossible to think of anything or anyone else. She remembered how she'd sit in her French lover's empty flat all day, staring at the wall, paralyzed, waiting for him to come home so she could come back to life.

"Is that all right?" he asked, and Jamila realized he'd been speaking, he might have repeated the question several times already. They agreed to meet the next evening. He'd wanted to come over immediately—bored? overeager?—but Jamila's brain felt too overloaded to deal with him.

Her marriage was progressing at great speed. It seemed to be on automatic pilot now, with her mother and his racing around India and Pakistan making preparations. So, unless she was reminded by her fiancé's phone messages, which gave her a jolt of adrenaline and which she alternatively returned or erased, she managed to forget about it most of the time.

Except at night. Then the dreams returned with such fervor it was impossible to ignore them. Now her skin was being peeled off her still-breathing body, like an animal's at the tanner's, a slit like a zipper cut from her lower lip to her crotch. Her organs spilling out, her heart pumping blood that spurted uncontained in all directions. Now she was being shoved into a tiny space, padded and airless, a suitcase. Her back, her neck ached from being folded for so long. The last trickle of light and air disappearing as the buckles were snapped shut. In the mornings, under the shower, she'd feel more exhausted than she had eight hours earlier. The refreshing shower gel, the energizing massage glove, the revitalizing shampoo, even the dynamic toothpaste—nothing made a difference. So she agreed to have dinner with Pierre.

They went to a magazine party first—all fashion models looking beautiful and vacuous and editors and writers looking intellectual and striking, their awkwardness played up like a gleaming accessory. For someone who fancied herself one of the intellectual set, Pierre was the perfect foil. Jamila opened her front door and realized that he was handsome. What had first read as ordinary now read as classic. Mediocre to the point of symmetry. He was as handsome as a television father: blond, fit, clean white teeth, and clear blue eyes that seemed to reflect a purity of thought and decency of action. He wore a smart blazer and not too faded jeans. Best of all, when he spoke, it was in halting English, young and uncertain and sweetened with French. As a couple, she thought they looked great—she was intelligent and exotic and he was mainstream and unprovocative and he also took up the slack in the looks department. She loved the long stares he got as he walked away with her.

Afterward, he kissed her good night. A real kiss this time, slow and slippery. He dragged his fingers up the back of her neck and played with her hair, while the other hand seemed to pull her into him from the small of her back. However, it was the age of AIDS. They stopped at the same time. Neither of them was grasping and groping in the old way, stretching out every last instant into a painfully sweet ending. These days, everyone restrained themselves neatly. And mutually. It was not she, drawing the line of chastity and patience. Nor was it he, admitting to some past or present relationship or guilt over some manipulation.

Jamila slipped in the door and listened to it locking behind her. Pierre had sauntered away as well. The age of self-obstruction, she thought.

In the bathroom, she let her mind linger on his tongue and lips. She enjoyed the flush in her face and the lipstick smudged around her mouth as if she were a child who had gorged herself on red candy. The image pleased her. She giggled to herself as she brushed her teeth and washed her face. She hummed the latest song on the radio.

Jamila put herself to bed, luxuriating in the memories of the evening, letting her fantasies roll out around her like carpets, before gently sending them all skittering back.

That night, she slept like the proverbial baby.

The next morning, she was awakened by the telephone. It was her mother. She was calling, all strangled and crackly, from India, with a million questions about her favorite colors and styles and the sort of jewelry she'd like to wear. Her mother told her how wonderful her fiancé's family was—how cultured and refined. "You'll love them, Jamila, the house is full of art and books and music. They already feel like our family." Jamila also stood to gain a great deal of very beautiful family jewelry from her in-laws.

Her wedding clothes sounded unreal. All the preparations for the wedding heralded a secure and comfortable future. Jamila felt transformed for the second time in twenty-four hours. She was going to be like Cinderella. Lifted out of her cockroach-infested matchbox-sized apartment on the Bowery into a material dreamland. She was going to be an Indian Barbie doll in a pink silk sari, framed by delicate gold embroidery and sparkling rubies and pearls. She leapt onto the waves of her mother's excitement and was carried out to sea.

When she got to her office, the first thing she did was telephone her fiancé. She blurted out the plots of all her nightmares. She apologized for not having spoken to him much lately. She tried to squeeze all her fears into words and stuff them into the telephone receiver. She spoke to her fantasy husband. Finally, reduced to a tiny, trembling mouse, she dropped herself in his hands.

His reaction was not what she'd expected. Rather than being reassuring, he was irritated by her outburst. He told her he was disappointed. She was the one who wanted an arranged marriage, wasn't she? And now she was fighting it the whole way. As if she was being forced into something. Which she was not. He told her how great it was going to be as soon as she calmed down. He told her all his plans for their future. He had a promotion coming up. He was climbing higher up the tennis pyramid. He was buying new carpeting for the bedroom and he wondered if she had any preferences. He thought she should renew her driver's license and start practicing now, because then their car insurance would be cheaper. Did she know what kind of a car she wanted?

Jamila flashed back to the night in his sparkling kitchen. When she looked down, she'd been blinded by the glare of his polished floor against her patent-leather pumps. He pulled a jar of black lumpfish eggs out of the refrigerator, pleased that he'd remembered that she liked caviar. She was horrified that he hadn't gotten beluga. She couldn't bring herself to explain what a sturgeon was. Instead, she looked back at the floor and noticed a discolored stripe. With his smug smile, he told her he'd spilled a jar of capers. She was confused by his pleasure. He said she should be impressed, because capers meant smoked salmon. Without thinking, she answered that capers meant spaghetti sauce. Instantly, she was ashamed of her snobbery. How could she be so unkind? As she hung up the phone, she knew that what she really wanted was to see some crack in his perfect shell.

She spent the rest of the day at work trying not to think about him. She typed a letter for her boss thanking someone for something, just a professional courtesy—no, an obligation—rather than from any deeper sensation. Her boss had an amazing distance from his emotions and looked at sympathy for others as a weakness. When he looked at others at all. Needless to say, he was not softened when he heard of her current situation. His main concern was whether or not she could find and train a replacement before she disappeared into conjugal limbo. When Jamila described her boss to her fiancé, he laughed and said, "He's obviously very highly evolved." Jamila was not amused. Fortunately, the onslaught of her boss's constant demands made sure that she thought of nothing else.

It was ten minutes to six before she looked at the clock again. She was throwing stuff in her backpack when the buzzer went off. She had an instant of panic wondering if she'd forgotten to alert her boss about an evening appointment. She tried to just ignore it and pretend that it was some bum on the sidewalk. But then it rang again. Louder and more insistently. Her boss shouted, "Who is that?" in an accusatory tone.

She pressed the talk button. "Hello?"

"C'est moi."

The intercom was badly designed, so most of what she heard were the trucks going past on the street below.

"I thought you said I didn't have anything for tonight!" her boss said from his office.

"Hello?" she tried again.

"*C'est moi, Pierre.*"

And a few seconds later, he was walking into the office. For a second time, Jamila didn't recognize him. He was dressed in the style of a young television lawyer, in a chocolate Armani suit and small wire-rimmed glasses. "Hi," she said nonchalantly. It occurred to her that she did not know what he did for a living. In fact, she couldn't even bring his surname to mind. Her boss started wandering out with his briefcase, but stopped short when he saw Pierre. Pierre's suit seemed to imply that he was waiting for the boss.

"This is my friend Pierre," Jamila explained. Being in the media, her boss was easily impressed by appearances.

"Hi," said her boss with unusual friendliness. He shook Pierre's hand. "Good to meet you. See you soon. Did you have my car brought down?" The question was addressed to Jamila.

When he'd left, Jamila turned to see what Pierre thought.

Pierre was looking in the mirror and straightening his tie. "You like this?"

"Sure," said Jamila. She was actually thrilled to see him looking so respectable.

"It's not very good," said Pierre. "I can't do things in it. But it makes me look—uh, intelligent, no?" He fluffed his hair a little. "I was working today. Tomorrow, I'll do it again. So I keep the suit. I look handsome?"

"What work?"

"I am mannequin. Mo-del." He pulled a heavy black notebook out of his bag. Jamila opened it. It was full of photographs of Pierre, torn from the pages of magazines. Pierre in tennis shorts, returning a serve. Pierre spoon-feeding ice cream to a lovely young girl. Pierre standing at a business meeting, showing some charts. Pierre at a desk, looking seriously into a computer screen. Pierre in a bathing suit, poised to dive into the glassy surface of a blue swimming pool. Pierre, as she expected, teaching a little boy how to ride a bicycle. The model boyfriend, father, husband. "You are . . . *secrétaire?*" he asked.

"Of course not," Jamila answered, offended. "I'm an editorial assistant. *Editeur.*"

"*Editrice!*" He laughed cheerfully. "Dan. You know Dan? The blond. He told me where you worked. I'm working nearby. So when I finished, I decided to see you."

As he walked her home, she said, "You look so ordinary. I mean, normal. I never would have guessed—how weird that you're a model."

"I prefer not to tell people that. They think . . . models are stupid. But I would make a good actor, no?"

Jamila never would have admitted it, but she was impressed and intimidated by Pierre's book. Model. These days, just the word conjured up fairy tales, images of money, glamour, and luxury. And she had a hard time not believing the advertising. All of a sudden, she noticed the looks he got from people on the street. Packs of teenage girls stared hard at him. Sophisticated women drank him in with their eyes. Gay men swallowed and stared with so much long-ing she thought he might consider changing his orientation.

If asked, she could act blasé and say: Oh God, a model, what a dimwit, all brawn and no brains. But in real life, she kept seeing herself as his match in each of the photographs. Then again, why would a model be interested in her? He was probably constantly surrounded by fresh beauties, barely in their teens. In an instant, he changed from a pastime to something worth owning. "Sure, you'd be a great actor," she reassured him.

They slept together that night. The novelty made it exciting, but he lacked a certain spark. It was pleasant. Not thrilling. As it turned out, his welcome was wearing thin where he was staying anyway. She told him he could stay with her, for a little while, if he needed to. The only thing she asked was that he not answer the phone.

The next morning, he got up early with her. He made very strong coffee for both of them and came with her up to the swimming pool. He watched from the viewing deck while she swam her lengths and then gave her tips on improving her technique. He accompanied her to her office and kissed her goodbye at the door. He even telephoned a couple of times during the day to tell her how the shoot was going.

At exactly six in the evening, he buzzed to collect her. A few days later, they were settled into the routine of a married couple. Occasionally he didn't have work or worked someplace else, so that he didn't meet her at the office. But he came to the pool with her every morning and said "Bravo!" when she got out. And he always telephoned a few times during the day to say hello. Jamila was pleased to come home and find him naked to the waist, wielding a wrench or a screwdriver, doing touch-ups on her apartment, looking like a joke rendering of an Adonis in modern-day life. His golden chest was damp with perspiration and his blue eyes glowed like sapphires with his effort. Jamila was increasingly aware of his beauty.

He also cooked buttery three-course French dinners for them—all she had to provide was money for groceries—and lit the candles on the table every night. They talked about philosophy—like most French men, he was enamored of Hindu mythology, the *Kamasutra*, and one Indian guru or another—about his acting career, about what he needed to do to make it work. She didn't talk to him much about her job because (she would have said it was too difficult to translate into French) she didn't want to focus on the fact that she really was a glorified secretary, after all.

She was well fed. Every hinge, electrical outlet, window, and drainpipe in her apartment worked perfectly. He scrubbed the bathroom. He did laundry. He arranged the spices in her cupboard. He put her shoes in rows of matched pairs. He found the cases for all the tapes and CDs and put them in alphabetical order. He traveled with only his backpack, so the place was never littered with his clothes and possessions. Jamila was not bowled over by his conversation, but she enjoyed the company. She loved the romance of doing it in French. She had been hypnotized once and told she was highly susceptible to suggestion. Her ability to envision the Pierre of his tear sheets animated their otherwise ordinary sex life. His muscular body and regular breathing in her bed also tamed her subconscious into submission. She slept long and well.

One evening, she came home to a rapidly blinking answering machine. She was jerked back into reality by messages from her father, her mother, and her fiancé, all discussing the plans for her marriage. No one managed to reach her at work, and they all wanted answers immediately. While Pierre sautéed something delicious, she called her father. She told him she'd been busy lately with some new friends, but she was fine, didn't need anything. The reason she hadn't rung back her fiancé was that she wasn't sure about what she was doing. Whatever you do, warned her father, don't tell your mother that, because she's all the way in India and she'll get worried for no reason. He reassured her that her fiancé was a very good choice.

Next, she tried to ring her mother and got only servants, who

C 142

were unclear about her whereabouts and with whom she left messages in her pidgin Hindi. Finally, she rang her fiancé. He was feeling a bit romantic. He wanted to wax lyrical about her independence. He hoped she'd be back home soon so that he could take her to restaurants. He loved eating out and knew she had refined tastes. He wanted to take her to all the best places in town. He wanted to hear what she thought. Also, he loved buying art and he thought he should get her opinion before he bought anything else, especially since she hung around with an arty crowd.

She was cool and evasive on her end, wondering all along if Pierre was listening. She kept thinking that it would be better if he heard it from her than surmised it secondhand. However, when she hung up, he seemed his usual cheerful self.

Even then, after dinner, she sat him down solemnly and told him of her upcoming marriage. She expected him to be upset, or at least a little hurt, but he seemed unperturbed. He thought getting married was a good idea. And being a nice girl from a good bourgeois family, she should marry someone from her own class. He, on the other hand, came from a very poor Polish family. His father was a mechanic and barely spoke French. His mother worked in a sweatshop. He told her that as a child he'd had to steal apples from his school lunch to give to his little brothers and sisters for supper. He had a girlfriend, too, in Paris. An older woman, a secretary now, with the same kind of childhood he had had. She'd started out as a prostitute in the Bois de Boulogne. She had a full-grown son almost his age. They were soul mates, Pierre told her. When he went back to Paris next week, he'd have to bring her gifts of lingerie and perfume to placate her.

Jamila was shocked. Not least because of his background. She realized she didn't know anyone poor. Not really. And prostitutes in the Bois de Boulogne! She remembered looking at them like zoo animals through the rolled-up windows of a friend's car, driving away quickly if they dared approach.

While psyching herself up to tell him, Jamila had entertained fantasies of Pierre saying he could never share her with anyone else. He'd insist she break the engagement immediately. He'd ask Jamila to marry him. He'd work hard to show her family that he could be respectable. He'd ask for her help in learning to be upper-class. She'd envisioned herself being the tearful voice of reason, telling him sadly why it would never work. With blurring vision, she'd explain that she'd been promised to another against her will, but that, to preserve her family honor, she had to do it. She ended her fantasy with a tragic Romeo and Juliet/Pyramus and Thisbe style of conclusion.

Instead, Pierre congratulated her and finished washing the dishes. Jamila was incensed. She decided that he'd better fall in love with her so that she could drop him properly.

That night, Pierre kissed her gently on the cheek. He brushed the hair off her forehead. He pulled her into him across the mattress, murmuring French nothings. "My little chicken, my little cabbage . . ." Then he drifted off into the comfortable sleep of a long-time companion. Jamila tossed and turned and coughed, but failed to regain his attention. Finally, she spoke, very softly. "Pierre?" She put on her sexiest French accent. She kissed his eyes, his nose, his ears. At the same time, remembering the advice of an old boyfriend, she stroked his entire body with both hands. While one hand traced feathery circles on the inside of his thighs, the other pinched his nipples. "Pierre, you're not sleeping, are you?"

"Mmm."

"Pierre," she murmured, "do you have any fantasies?"

He rolled over.

"Tonight, I'm still yours, Pierre." She licked his earlobes. "We can do anything you want . . ."

"There's something I like," he said sleepily, "but you could be—surprised."

"Surprise me."

Not surprisingly, he wanted to see her collection of lingerie. He wanted her to wear lacy teddies and stockings with a garter belt. He wanted to see very high heels, sparkly earrings, and red lipstick. He knew the brand names of all the best manufacturers of intimate apparel, makeup, shoes, costume jewelry, knew every style and color—"Do you have one like this? La Perla makes them"—better than the most avid fashion magazine reader. Their love life gained momentum. The usually brief interlude and loss of consciousness stretched into an endless marathon of sweaty half darkness and twisted bedsheets. Jamila awoke with a sour taste in her mouth and her eyes swollen and sticky.

At the swimming pool, she was wiped out. Everyone kept swimming around her in the lane. She was left choking in the splashing wake of someone else's kicking feet. Finally, she was forced to make the humiliating bob under the rope from the fast lane to the medium. Pierre sat on the bench, watching her greedily as she emerged.

When they reached her office, he kissed her with passionate abandon. Then he whispered, "Would you like to try something else?"

"Of course," she said. "Anything you want."

"It's something very sexy," he growled. "You are ready?"

"Anything," she said confidently. She was beginning to have second thoughts.

"Until tonight," he said, suddenly sounding like Pierre Brel.

Jamila walked into the elevator thinking that he looked older than she thought he did when they first met. In the office, her boss growled that he'd been answering her phone calls all morning. She gave him a silly grin and asked him if he'd taken messages. Then she shambled, like a happy drunk, to her desk. There was a stack of pink while-you-were-out sheets. She pushed them aside and started doing her boss's expense reports. Mindless work. Entering figures into the computer was all she was capable of. And she was so blissed out that she wouldn't get depressed at his entertainment expenditures.

But a half hour later, her eyes wandered. The top of the stack was her mother, returning her call, from India. She pushed it aside. She could hardly be screaming into the receiver in Hindi while her boss closed million-dollar deals. The next one was from her fiancé. He wanted to know where she'd be that afternoon. There was one from her father, asking if she was feeling better. Then there was one from a business of some sort, confirming her home address. Her father again, wishing her Happy Valentine's Day. Valentine's Day! She looked at the calendar. Had she been engaged for almost two months? She reminded herself to send a belated card to her fiancé, just pretend it had got lost in the mail or something. Then she drifted off into a reverie of the previous evening.

When she got home that evening, the sidewalk was blocked with refrigerated vans. There was a knife-sharp icy wind blowing and her frozen fingers could barely get the keys out of her backpack when she realized she wouldn't be able to get in the door. Cursing, she had to walk all the way around the block and come back the other way. It was probably one of those damn filmmakers, she thought, using the city like a giant studio as usual. The bums had spilled out of the men's shelter onto the Bowery to watch all the

commotion. Some of her dissolute neighbors had, even in this brutally cold weather, screeched open their windows and stuck their
heads out to shout obscenities at the workers.

When Jamila walked into the lobby, a couple of delivery guys
squeezed into the elevator with her, stopping a few times to reorga-
nize their overpacked dollies. They were delivering tropical plants
of some kind. When the doors finally slammed shut, the elevator
was filled with the sweet scent of flowers and leaves. Jamila closed
her eyes and inhaled. How gorgeous. Their damp coats steamed, a
fragrant humidity. For a few seconds, she disappeared. She was in
India with her mother. They were watching the sun set in the garden, surrounded by gardenia, bougainvillea, and queen of the night,
luscious scent and abundant color. A servant was bringing a tray of
refreshments, walking toward them in the purple light.

The elevator door opened with a cold draft, and Jamila said,
"Excuse me, this is my floor . . ." And they all struggled to get out
at once. Jamila rang her doorbell. The deliverymen laughed, "Hey!
These for you? Whadja get married?" The deliverymen pushed past
her inside, unloading the packages wherever they could and tearing
off the wrapping. They were flowers. And flowering plants. Huge
clusters of brilliantly colored tropical flowers. Everywhere Jamila
looked, there were flowers. Leaves and flowers. The apartment
looked like the rain forest.

"Be right back," said the deliverymen.

"How amazing . . ." said Jamila.

"Is the third trip," mumbled Pierre, looking uncharacteristi-
cally sheepish. He told her that he tried to help the guys, but they
said they were fine on their own. "Not very friendly," he said. In
New York, Frenchmen were not recognized for their brawn.

A few minutes later, they returned with another load of flowers. It was lucky that was the last one, as no surface space remained in the apartment. They handed her the receipt to sign. "Congratulations," they said as they left, glaring suspiciously at Pierre.

"Who sent them?" she asked, looking hopefully at him.

He pointed at the card stapled to the last bouquet, on the table.

Jamila tore open the envelope. It was inscribed:

To my (future) wife,
the first of many
happy Valentine's Days . . .
with love.

Jamila was sick to her stomach. She crumpled it up and threw it in the trash. "They're from my fiancé," she said, thinking that such a large display of affection could not fail to arouse some feeling of possessiveness (or even competition) in Pierre. It did not.

They ate dinner engulfed in hothouse flowers, airlifted from Singapore and Bangkok. Orchids brushed her cheeks as she chewed. Jasmine crept down her ears. Leaves and petals sprinkled the plates. Suddenly the flowers seemed frighteningly artificial to Jamila. They seemed to smell of chemicals and pesticides. The leaves gleamed, waxed and polished. The roses stood on gracefully slim shafts, shorn of thorns and imperfections.

"Very beautiful," said Pierre.

Jamila pulled a rose from the vase. "Here"—she dropped it on his plate—"have one. Have ten if you want. Have them all." She waited.

Pierre simply looked pleased. "Thank you."

"I'm going to bed," said Jamila with a sigh of resignation. It occurred to her that Pierre's obliqueness was not a result of the language barrier. And she was fed up. She disappeared into the bedroom. While changing her clothes, guilt overwhelmed her and she decided to call her fiancé. She knew he wouldn't be home yet, so she called his machine. She preferred that to an awkward conversation with him at work. "I just got home to the most lovely surprise!" she said, with as much enthusiasm as she could fake. "It was more incredible because I didn't even realize it was Valentine's Day!" Oh-oh, she'd blown her card-got-lost-in-the-mail routine. "I've never had so many flowers in my life." That was the truth. In fact, she was sure few people did. Unless it was at their funerals. She collapsed on the bed.

Pierre poked his head in the door. "Jamila?" She acted as if she was sleeping. "Jamila, do you remember . . ."

Jamila opened her eyes. Her bedroom was a mess. She didn't remember leaving it like that. "What a mess," she grumbled.

"Do you remember," he said, hesitantly, "we were going to try something . . ."

"Oh sure," she said as it came back to her. "Anything you want."

He walked back into the bedroom carrying his bag. He threw it on the bed and pulled out his model books. Then his gym shoes. Then a towel and a sweatshirt. Reaching the bottom of the bag, he pulled out a handful of black lace and a pair of black high-heeled pumps. He looked at her, waiting for a reaction.

"They look like the ones I wore last night . . ." she said as it slowly dawned on her that they were for him, not for her. Man-sized shoes and lacy stockings and a bustier. She looked back at

Pierre. His face was red and he was breathing hard—from embarrassment or excitement?

"Do you like them?" he asked, leaping up and pulling off his clothes. In seconds, he had the bustier over his head. Then he was sitting on the edge of the bed, the seams perfectly lined up on his stockings, as he slipped his feet into the pumps. Jamila felt like giggling hysterically. But she was hypnotized by the strange fury with which Pierre dressed himself. "Do I look beautiful?" he asked breathlessly. He walked back and forth across the floor, sliding his hands up and down his bottom, moving in an exaggeration of a sexy woman, like an exotic dancer. "Do you like the way I look?" He rushed over to her dressing table and started applying vermilion lipstick to his lips. "Am I sexy?" he whispered, more to himself than to her, she thought. He stood up again. Stood in front of her. Started swaying in a slow, seductive dance, "Look at my ass. Isn't it sexy?" He turned around. "Look at my legs, my long, long legs, aren't they sexy?" A caricature of a hooker.

Jamila exhaled. Sexy? For a second, it was a joke. Then she was sucked in. Compelled. He caressed himself. Strangely enough, it *was* sexy. Rather, his trancelike state, his fury was. He believed it. He felt it. He was so excited by the game, the ritual, that he was transformed. Like a Method actor. Or a religious zealot. "Tell me . . ." he growled. Not half man, half woman, but a man's ideal woman. The inflatable doll brought to life. The woman he wanted.

When Jamila thought back to that night, the last few nights before Pierre went back to Paris, she remembered them as though lit by neon lights. The room seemed to change from pink to blue to purple and back again. Like the light from a truck-stop sign. All the while, Pierre's voice barked instructions, she was a man, a man looking for a good lay, a prostitute. Was he what she was looking for?

First he was the man, then he was the woman, then the man, all in a strange manic frenzy. She watched, fascinated, played the roles assigned to her like a puppet.

That morning, Pierre woke in a dark, brooding mood. He made an unusually bitter brew of coffee, so opaque she could see her reflection in the surface of the liquid, but undrinkable for anyone but him. He accompanied her silently to the swimming pool. He watched impassively as she swam. How strange, thought Jamila, when despite the lack of sleep she found the presence of mind to think. But she was pleased when he dropped her off and kissed her goodbye, standing in a crunchy drift of snow.

All of a sudden, he reappeared from his shell. He looked pale and exhausted. But he looked at her. For the first time in hours. As though he could really see her. He held her face in his hands and brought it carefully to his lips, as though she was made of glass. Very precious and fragile. Then he touched her eyelids, so gingerly she could barely feel it. He kissed her again before he walked away.

Her fiancé rang the moment she walked into the office. He was angry that she hadn't bothered to call him at work. He was coming to see her, he said, they had a lot to talk about. The wedding was now less than a month away and there was so much they didn't know about each other. So much, thought Jamila. He said he'd be there the next weekend. Fortunately, thought Jamila, Pierre would be back in France by then. Jamila apologized profusely. The apologies were not accepted. However, he did seem cheered up. He told her he'd booked a hotel in Bali for their honeymoon. He'd read that Bali was the only place to be this resort season.

Her boss had left a magazine sitting on her desk, with a marker sticking out. She flipped it open. In it, an attractive couple lingered on the beach and then switched to formal wear for the evening. The man was Pierre. Her boss hadn't written a note.

When she asked Pierre about the pictures, he told her the shoot had been great. The resort was fabulous. He lived on grilled fish and coconut water. They'd danced and drank local potions all night. Unfortunately, male models were never as highly rated as women and he'd had to stay in the staff quarters. In the end, though, he hadn't minded that, because they were the kind of people he'd rather hang out with anyway. He said he wished he could afford to go back someday.

As they got ready that night, he froze midway. "You don't think I am a transvestite, do you?" he demanded. "You don't think I am a homosexual?"

"Of course not," Jamila reassured him. Stroking his cheek, she admitted—to herself—that she didn't know what she thought, but it was all incredibly bizarre.

"No?" Eventually, he relaxed again.

The phone rang at about three that morning. Jamila could barely stand to look at the clock; even its muted light seemed to pierce her eyeballs. But the phone kept ringing, she must have forgotten to turn on the machine. She went to the living room to pick it up. It was her fiancé. He was drunk. He was calling to tell her that he knew how wild she was. He knew she was going to teach him how to really live. He couldn't wait until they were married. He remembered the night they kissed. He thought about it a thousand times. He remembered holding hands with her at the ballet. They'd kissed and touched each other. He thought about her all night. He sounded desperately innocent. "I love you," he said. "You love me . . . ?" A barely audible whisper. He was so soft she could have sliced him up with a Q-tip. Jamila pulled the blanket over her shoulders. She fell asleep on the sofa.

But the second her eyes closed, her demons returned and chased her into the morning light.

The next night, the marathon crashed to a close. She was so tired she could barely walk. Let alone have a conversation. She didn't even eat any of Pierre's dinner. She simply fell face down on the bed, fully dressed. When she woke up, uncomfortable and hot, it was four in the morning. Pierre was cuddled up in the duvet, breathing deep breaths of sleep. Looking at him, she remembered his sheepish expression when the night deliverymen came. She remembered the way they glared at him. Remembered what a mess her room had been.

Fortunately for Pierre, it was the night before he was leaving for Paris. Because the next thing he knew, Jamila was shaking him awake. "Goodbye, Pierre. You have to leave now." She ran around the room, throwing his things into his backpack. He'd once told her that he didn't mind sleeping on the E train. It was clean and well heated. He could do that tonight. "I'm about to get married," she said. He looked at her blankly. He sat up and got into his jeans.

"Jamila."

Great accent, she thought.

"I will call you when I am back," he said. He zipped up his nylon coat.

"Sure," she said as he walked out the door. "Have a nice trip."

She fell asleep again. A deep sleep. Almost a coma. So sound that neither her nightmares nor her alarm clock could intrude. When she awoke, it was past ten. She called the office and her boss was fuming. She promised to be there soon, but she didn't rush. She made her own coffee, comfortingly weak. She browsed through a travel magazine. She got dressed in the clothing that most closely resembled her pajamas.

At work, her fiancé called, embarrassed. It had just been too long since they'd seen each other. Everything seemed unreal. She agreed. "It's all right," she said. "But I don't love you." There was a crevasse of silence. She looked at the diamond engagement ring. "This is an arranged marriage," she said. "It's not about love. Or sex. If you don't understand that, we shouldn't do it." He apologized again. His voice quavered and broke. Everything would be all right when he was there, he said, his words wobbly and small. She didn't answer. Then he started to cry softly. In a voice thick with tears, he pleaded with her to withhold judgment until he arrived. Between sniffs, he promised her they wouldn't talk on the phone for a few days.

"Is there someone else?" he asked, and she said they'd talk when he got there.

She opened an envelope on her desk. Her father had sent her a photograph of their wedding day. Her mother had been a very shy bride of sixteen. Like most Indian girls at the time, she hadn't been given much choice in the matter. But her father looked equally shy and awkward. He'd gotten married to her on his family's advice. The colors in the photograph had begun to fade. The rich reds of her mother's wedding sari had turned pale and greenish. Her father's skin had a sallow cast against the wilted jasmine flowers. And there were two wedding invitations. One was stained and slightly bent, with her mother's and father's names engraved in the spidery nuptial style of the time. The other was crisp and new, intimidatingly heavy, with a tiny angel in the corner. Jamila's name was first.

She imagined herself, in ten years, telling her children about their father. Then, in thirty years. Imagined herself in her mother's position. She'd kept meaning to ask her mother if she was happy. She didn't because she thought maybe her mother wouldn't know the answer, wouldn't have ever had anything to compare it with. "Life is not just about making yourself happy," her mother had said to her when she was seven. She had said she'd be happiest if she just had lots of dogs and horses. She wanted to be a farmer. Alone with the animals. "You have responsibilities to a lot of people," she'd said. "You're not all alone out there."

Her boss wandered out of the office. Her presence had cheered him up. "Do you want a cappuccino?" he asked her. "How're the marriage plans going?"

"Great," she said. "All under control." As he got on the elevator, the front door buzzer went off. It buzzed again. Jamila ignored it. She was sure she'd sleep well tonight.

Lynne Sharon Schwartz

$\mathcal{A}$ C Q U A I N T E D

W I T H   T H E   N I G H T

Alexander Smith woke to find himself sitting up in bed. The bedside wall lamp was on. His glasses were still on, and a book lay open on the blanket, two middle pages peculiarly upright, swaying in the faint fall breeze from the nearby window. The digital clock said the time was 2:47. Odd how the last two numbers were his age, a reminder in the dead of night. "Shit," he muttered. He hated to doze off reading, which had been happening to him about twice a week lately. He had trouble getting back to sleep, and mornings after, felt jolted out of sequence, as if two days had passed instead of one. A small click sounded; it was 2:48. There, he had aged. Staring at the straight-edged, unfriendly numbers, he vaguely recalled a Robert Frost poem that said a solitary clock proclaimed the time was neither wrong nor right. Yes. That was exactly how it felt in the middle of the night. The time was just a meaningless number with no attachment to events. Alexander felt stranded and forlorn.

He put aside his glasses, switched off the lamp, and got down under the covers. He felt the warm back of his wife, Linda, turned away from him, her contours familiar and soothing. He hadn't thought to look, when the light was on, to see if she was there. But where else would she be at two forty-eight? A click; he made the correction, forty-nine. Just so, it went by.

Sleep eluded him. To make matters worse, he discovered something quite strange nagging at him. A small shape, dark, yet standing out against the deeper dark, danced behind his closed lids. It looked like a bacillus. Alexander's eyes had been strained lately, since he used them constantly in his work as an architect. Perhaps he needed stronger glasses. Perhaps he was getting old. Undoubtedly he was getting old. He watched the bacillus dance about, and found that if he rolled his eyes from left to right the spot moved with them. If he rolled them up and down the spot went along too, but the up-and-down motion hurt.

Minutes clicked by, and it would not go away. It made his skin tingly and restless, as if his insides were struggling to escape from their container. He knew what it was, though, and knowledge was reassuring. The spot was an aftereffect of sleeping for two hours with the light on and then waking up to the dim glow and going abruptly back into darkness. It was some optical phenomenon he couldn't explain precisely, but whose broad outlines he felt he understood. As a matter of fact, he felt that general imprecise understanding about a great many things, he realized: the tides, rocket ships, airplanes, rainbows. Maybe he really didn't know anything thoroughly. What the hell, though. He managed, didn't he? Now sleep.

Alexander opened his eyes in the dark. He could see nothing. It was too soon. You had to lie awake for a while in the dark before you could see everything. He saw only the spot, dark against dark, floating through the void like a flying saucer. No longer shaped like a bacillus, it was a small circle with undefined edges, rather like a planet seen through a telescope, with a halo around it. Or a gray star with a gray glow. He closed his eyes; it remained, spinning, creating a haze, a wake of its motion. Horrible. He opened them. He could begin to distinguish the furniture now. The room was spacious. There was his armchair against the far wall. Then his bureau drawer on the right; Linda's was on the left. Above Linda's was a mirror, illumined in places where moonlight glimmered in through the window. The spot went everywhere Alexander's eyes went, relentless. It flickered in the jagged beams in the mirror. He couldn't get rid of it. A UFO with a message. Glaucoma. Retinitis pigmentosa. Impending death, beckoning. To ease the panic he moved closer to the warm body of Linda. She was wearing a thin silky nightgown that excited him mildly as its smoothness brushed against his chest and thighs.

He realized he was trembling with fear. Maybe he ought to make love to Linda. That would at least be something to do while he couldn't sleep. She was still turned away from him. He put his arm around her and pulled her closer, testing the strength of his desire. It was nice making love to Linda. He pressed against her. She was usually an eager partner, and if not always totally eager, if some vague, ancient tug seemed to hold her back, she was at least amenable. He put a hand on Linda's breast and eased a knee between her thighs. The spot in his eye throbbed, zoomed forward and back to tease him, taunt him, like the cavorting spot at the end of an invisible laser beam. Did he want to make love to Linda? He queried his body. Actually not very much. He was tired and distressed by the frustrating day and longed to sleep.

But maybe he should do it anyway. It might make him forget about the thing in his eye. Once he started he would want to. He moved his palm around Linda's nipple but she did not stir. God, what a deep sleep! He envied her. His eyes rolled involuntarily with the motion of his hand, and he noticed that the speck rolled too. It was terrifying. Trembling, he turned over on his other side, leaving Linda. The clock said 3:04. The right-hand numbers of the clock stopped at sixty. Maybe he would die at sixty. Or the next sixty. Actually they stopped at fifty-nine. There was no 3:60. The spot was on the clock, on the upper-left-hand tip of the four. Five.

Alexander began to experiment with the spot. If he could not get rid of it he could at least play with it, tease it back. He stretched out flat and looked up at the ceiling. He blinked. The spot disappeared for the fraction of an instant that his lids fluttered down and up, but immediately reappeared to jiggle on the ceiling. He began to blink to the rhythm of the first movement of Beethoven's Seventh Symphony and the speck obediently danced. But the fast pace made his eyes ache, so he lowered his lids to rest them. He didn't feel like playing with it. He was exhausted. There was no comfortable way to arrange his body; his pores seemed about to burst open. The speck was an intrusion, undeserved, unbearable. He wanted to cry out in protest, as he would protest to the police if a thief entered his house, but there was no one to protest to. It was his very own speck. He thrashed around in the bed, viciously kicking the covers about him. Then he pressed his fists hard to his eyes and for a moment

found relief. Gone! But when he released them it was back, surrounded by colored flashing dots. They went away gradually but the
spot remained. Alexander started to sweat. He hated the spot sav-
agely. It was not in his body—it seemed located in distant space,
yet it controlled his existence like a vital organ, heart or lungs. Then
he quieted in surprise, for the way he had just described the spot,
distant yet part of him and controlling, sounded like the idea of
God that was taught to children. The speck was God. God was
paying him a nocturnal visit. A vision.

Alexander couldn't believe it was himself having such alien
thoughts. His brain was softening. Premature senility. He ought to
laugh; he must be delirious. But it was not funny. Very possibly this
was the way people went insane. God. Shit, he thought. He would
never read in bed again. His forehead was cold with damp sweat.
He went into the bathroom for a drink of water and looked in the
mirror, but couldn't really see himself because of the mote in his
eye. There was only a haggard, generalized familiar face: anyone's, a
good-looking model for expensive scotch in the pages of a slick
magazine, caught unawares in his pajamas, with a hangover. The
mote was in the mirror, on the pupil of his right eye. It was a gross
distortion of figure and ground. He was the ground and the mote
was the figure. Blood surged through him. Furious, he lifted his fist
in a violent gesture to smash the mirror, but stopped himself in
time. He really must get hold. Maybe he ought to read for a while.
But he knew that the speck would move along the words of the
page; he knew exactly how it would look, gray, bouncing along the
white page, a replica of his eye's movements, and he didn't want to
try.

Back in bed, he pressed the pillow hard over his eyes. Thank
God! It was gone. Maybe now he could sleep, if he could find the

right position. Soon it would be morning. He peered out: 3:58. The hour was aging. Linda lay calm; she hadn't stirred. Linda was forty-one. He experimented with the pillow; at last, lying on his stomach with his face pressed into it but turned slightly to one side, he could keep the mote away and still breathe.

Sleep did not come, but he was more peaceful. He tried to think of nothing, but events of the day, blueprints, drawings, the faces of his associates, ran through his mind. It had been a troubled day: a contract they believed they were sure to get was at the last moment given to a younger, rival firm. Alexander had been ruthless, shouting at his staff and threatening to fire people for not working hard enough. He had felt weighed down with the burden of the business pressing on the front of his head. He thought of women he had seen in Caribbean islands moving gracefully down dirt roads with huge, heavy baskets on their heads. They sailed along, proud and erect. He staggered, clumsy and in pain, beneath the burden. Now he could see that the contract was less important than it had seemed. Perspective. The firm was in no real danger. He shouldn't have carried on so. All right, so he had made a mistake. So he had behaved like a bastard; more like a frustrated child, actually. Was that a reason to be punished so harshly by this . . . thing? Nothing like this had ever happened to him before. He was a reasonable man, after all. But at least now he had it under control. He thought, in a while, that he might try a little test. Maybe the whole horrid episode was over, gone as mysteriously as it had come, and there was no longer any need to remain uncomfortably in this absurd position. So very gradually, as if afraid of being noticed, he raised his head from the pillow. Christ, it was still there! In a rage, he pounded his fist into the mattress. He wasn't going to get any sleep at all and he would be a wreck in the morning.

How could Linda lie there sleeping so calmly while he tossed in agony? It wasn't fair. She was his wife. She was supposed to share his pain.

"Linda." He shook her. "Linda," he called loudly in a hoarse voice. "Please," he added more softly.

"What?" She was still sleeping, he could tell. The word was a reflex.

"I can't sleep. . . . I have this . . ."

She rolled toward him. "What is it?"

"I have something in my eye."

"Go to sleep. It will go away."

"Linda, this is terrible. It's this thing. I can't stop seeing it. I can't sleep."

"Murine." Her eyes never opened.

"What?"

"Put in Murine. Drops. Bottom shelf."

Could she do all that in her sleep? Women were amazing.

"Listen to me. Wake up. It's not that kind of thing. It's something I keep seeing. I can't stand it."

There was no answer. She was sleeping. Alexander was enraged, but when he looked at the pretty curve of her shoulder he relented. What did he want from her? She hadn't sent the mote, and she certainly couldn't make it disappear. The mote rested on the peak of her shoulder's curve. He put his hand there; the mote was on the back of his hand.

"Alex," she murmured unexpectedly.

"What?"

"Hold me. I'm cold. And close the window."

Grumbling, he rose, shut the window, returned to bed, and held her. The spot was still now. His eyes were tired and not moving, and so the spot was still. It obeyed his eyes, a marionette of his eyes. Perhaps he was making a kind of peace with it. Perhaps he would have to live with it for the rest of his life. How would he manage that? He could spend the rest of his life staring directly in front of him, never moving his eyes, only his head. People would certainly think him odd. But seriously, he could get used to it. People got used to worse things. His brother wore a hearing aid. One of the junior partners at the office had had a toe amputated. And back when he was a boy, he recalled, before doctors became so adept at mastectomies, an elderly great-aunt of his had had a breast and a large part of her upper arm removed and had to wear her arm in a sling for the rest of her life. That must be very annoying. Of course, much worse than annoying, but for the purposes of this survey, annoying. Linda had a slight stammer when she got nervous. She knew just when it was going to happen, she told him. But she lived with it. It was a small thing, when you put it in perspective.

Thinking of all these things, Alexander was more wide awake than before. He detached himself from Linda, tucked the blanket around her, and rolled over, hugging himself tight. The mote was acting up again, bouncing back and forth like a Ping-Pong ball. He wasn't getting used to it at all. A person could get used to a sling—of course it was terrible to have cancer, but a sling was something you could accept after a while. You would be grateful simply to be alive. He would willingly wear his arm in a sling for the rest of his life if only this torment would go away and he could get some sleep. But that happened to women. Most likely he would have prostate cancer one of these days, he was nearing that age. Would he be so willing to give that up? Your mote or your balls? Wait a minute, I'll have to think that over. He remembered Esau, who sold his birthright for a mess of pottage. Ah, he understood now how men could make these foolish bargains. The speck winked at him; it was tiny, infinitesimal, a molecule. Maybe he was the only man on earth to have seen a molecule with the naked eye. It darted about wildly, flickered, floated, vibrated. He broke out in a sweat again, and yearned to die suddenly, right here, with no pain. He pushed his face in the pillow, but it stayed, even with the pillow. His last resort was gone. He wanted to cry from hopelessness. Maybe if he cried, some chemical reaction would take place in his eyes and it would go away. He rarely cried; all he could manage now were a few weak tears that had no effect.

All right. It was going to stay for a while. He would accept it. Be reasonable. He tried to lie still, though his skin stung with frustration. He would think it through. The mote must be more than a mote. It must be a symbol. It represented something about himself that he refused to face. That was how it worked: you buried something, and it came back to haunt you in strange ways. He loathed self-examination. He really didn't believe the meanings behind things mattered very much. Action mattered, not motives. He supposed he was rather obtuse that way, at least Linda said so. But women in general were better at that sort of thing; it was their upbringing.

Still, maybe there was some awful secret about himself that he didn't know. A friend of theirs, telling them the sad account of her recent divorce, had presented the theory that everyone had a secret, a secret secret they didn't even know themselves. Her husband's secret, she said bitterly, was that he hated women. "Ron is a latent homosexual," she murmured. He and Linda had been shocked. Alexander was willing to accept that Ron hated women, but did this make him gay? It didn't seem logical. "Linda"—he spoke quietly into the dark—"do I have a secret?" She slept on. He hadn't really meant to wake her.

Maybe he was gay too, Alexander thought. Anything was possible. He was even willing to accept it if his acceptance would make the spot go away. He thought of several men he knew and juxtaposed them alongside the idea of his own possible gayness, latent or otherwise. Nothing happened. Did he secretly crave their hands stroking his body? He tried to imagine it, and felt neither disgust nor excitement. Lack of interest. "Linda," he said again. No, he could not ask her that. She was sleeping, and if by chance she heard she would surely laugh or else think he had gone out of his mind. He dismissed the idea of latent homosexuality. What next?

He thought of all the bad things he had ever done in his life and never confessed to. Once he had accepted a three-thousand-dollar bribe and distorted some figures in order to get a contract. He would not do it again today, but eighteen years ago they were hungry. He did not think, even now, that it was so terrible. He had done an excellent piece of work on the job, better than anything the competing firms would have turned out. Alexander searched on. He had not paid enough attention to his parents in their old age. He had been very busy at the time, getting the business on its feet and raising the children with Linda. He had let the distance between him and his parents grow until when they died it was almost as if he were burying strangers and had buried his real parents long ago, little by little, without ceremony. Yes, all right, that was bad, but they were dead now, in any case. He was sorry. Before that, when he was in college, a girl he slept with three times begged him for money for an abortion. He was poor and gave her all he could get together, seventy-five dollars. He didn't think he was the father;

she had a reputation for sleeping around. For days she phoned him, weeping, begging for more money, afraid to tell her parents or anyone at school. Finally in disgust he shouted at her, "It's not even my fucking kid. Leave me alone and get it somewhere else. From what I hear you have plenty of contacts." Was that so bad? She had indeed found the money elsewhere. Alexander felt he had a pretty good case. He rubbed his eyes; the mote bloomed astonishingly, then retreated to its familiar size, a speck, a seed. He shouldn't have said, "From what I hear you have plenty of contacts." That was gratuitous. I'm sorry, he screamed inside. For Christ's sake, it was ages ago. He had two more children now, grown. He had raised and cared for two children. Wasn't that enough? Now go away. But it didn't go away. It bobbed, like the little white spots used to bob along the lyrics of songs flashed on the movie screen years ago.

Once when his daughter, Sandy, was five years old he flew into a rage because she defied him, refused to pick up her toys from the floor, and he hit her hard all over, face, shoulders, arms, back. He stopped himself when her screams finally penetrated to him. No one else was home. He was alone in the house and abusing a child. How could that happen? It was something he read about in the papers with horror. He stopped and cried, "Oh my God," and held her in his arms, weeping, and apologized. It was agonizing. That was the only time he ever hit her. But was that so terrible, in per-spective? Didn't many men hit their children, more often, and did they suffer for it fifteen years later? It was very unjust. Sandy proba-bly didn't even remember. He could ask her next time she was home from college, but he was sure she wouldn't remember.

He had slept with a number of women during his twenty-one years of marriage, mostly when Linda was in the final months of pregnancy and the early months of motherhood—that was understandable—but at other times too. He met them in bars or at parties, saw them once and never called again. He wasn't proud of it, but he had never thought it was so terrible. Nobody was hurt. Linda never knew. It was simply a need he had at the time, he didn't do it anymore. All right, the women were hurt, he admitted. They were decent women, not whores. He always said he would call them and never did. As he was leaving he would say, "I'll give you a call in a few days." An easy way of saying goodbye. The words made him wince. "I'll give you a call in a few days." All right, that was pretty bad. It made him feel pretty low, remembering. But on the other hand, he had to be fair to himself, he never hurt them. Physically, that is. He knew that lots of times when a man picked someone up in a bar and got her home he used the opportunity to do awful, cruel things, really atrocious things. He had the chance, but never did anything like that. He was very nice with them in bed. He was as nice as he would have been with his own wife. Didn't that count?

The spot did not relent, even as Alexander dredged up all these things from the past that he never thought about anymore. It swirled in zigzag patterns, mocking, torturing. His body ached from tossing. Confessing his sins was no help. He felt more depressed than guilty. What was the use of dwelling on his own ugliness? So he was a worm, all right. Wasn't everyone? He was no worse than the next man. If everyone were to confess everything we'd all be in jail. Adultery, he thought, is technically a crime in this state. Possibly he could go to jail for it. He could also go to jail for child abuse, accepting a bribe, abetting an abortion at a time when it was still illegal. When he was a kid he and his friends used to steal miniature cars from Woolworth's. Theft too? This was becoming absurd. He pictured himself in jail, wearing a striped suit and cap, lining up with a tin bowl for nauseating meals, hacking away with a pick in a rock quarry, and he gave a small laugh in the dark because his image of jail came from old James Cagney movies. The man in the next cell would tap out a message in code on the wall, asking what he was in for, and Alexander would tap back, "Adultery." The other inmates would laugh at him. Would the mote follow him to jail too? Sleepless nights on a hard cot, watching a speck dance on the concrete walls of his cell. It would feel pretty much the same as now, except the bed here was soft and the room was large and well furnished. Yes, he thought, he was a good provider. He had provided his own cell and not made himself a burden to the state.

God, he moaned, help me! When Jack and Sandy came home from college at Thanksgiving they would find their father a changed man, aged, weak, fragile, and delirious. His children. Tears leaped to his eyes. The mote shimmered. He crushed the pillow between his fists.

There was one thing he had neglected to mention. All right, all right. For a year he had been madly in love with a young woman named April. She was an art historian who worked at a museum, and he met her at the opening of a friend's show. He had remembered her afterward only because he thought it was a ridiculous name. Then he met her on the street. They had a drink. And so on. He came to adore the name. After they made love he would sing her to sleep with all the songs he could think of that had April in them. April in Paris. April in Portugal. April Showers. He hadn't seen her in six years. The month of April was still a torment to him, though, writing the date all the time. He usually wrote "4" instead. No, he really must not think about her or he would go mad. Just the thought of her name in the dark filled him with sudden craving.

Lord, what was a man, at the mercy of a name. He looked over at Linda. He could wake her. You miserable bastard, he told himself. If you could do that . . . Hadn't he done enough already?

He remembered how Linda had confronted him with it. It was all over him, how could she miss it? They talked for a long time at the kitchen table late at night, rationally, considerately, about what they could do and what this meant in their lives. There was an air of unreality over their talk.

"It's not that I don't love you," Alexander was repeating calmly after an hour. "Don't misunderstand. These things happen."

"Yes," she said. "I can understand that."

She got up, took a pair of shears from a drawer, and cut off a great hank of her long dark hair.

"Linda!" Alexander stood up.

"It's all right, it's all right, don't worry," she said. "Sit down."

She put the hair in an iron pot and poured vinegar over it. It smelled foul and ugly. Then she lit a wooden kitchen match and set fire to the hair. Alexander could not speak or move. Civilized English had left him. He felt they were living before civilization began. This was a primitive rite that made him paralyzed and mute with awe. The hair sparked and crackled and soon the kitchen was filled with a hideous acrid smell that brought him to his senses.

"Linda . . ."

"Shh." And she smiled with her lips closed, and blinked her glinting eyes at him. "When it's all done I'm going to eat it."

He leaped up and shook her hard by the shoulders. "Stop!" he yelled. Her head wobbled back and forth. She looked terrifying, with half of her hair tossing and the shorn side jagged. "Stop!" he yelled again. He grabbed the sizzling pot and ran cold water in it. Dark hair overflowed into the sink.

She sat down quietly at the table. "You're sending my beautiful hair down the drain."

"Shut up."

"You're going to clog the drain and we have no Drano."

He cleaned up the mess and threw it in the garbage. Then he sat at the table opposite her again.

"Don't look so alarmed, Alex. It was only a passing thing. Everyone's entitled to a little temporary insanity." She seemed fine now, except for the hair. She chuckled, and touched the shorn side of her head. "I guess I'll have to wear it short for a while. I've been trying to decide for weeks whether to cut it. Do you think I'd look good with bangs?"

"Linda, do you want to hit me or something? Go ahead. You have a right."

"No. I just want to ask you a question."

"Anything."

"What did you do with her?"

"What?"

"What did you do with her? You know. I mean, the same things you do with me? Tell me what. I want to imagine it."

"Linda, really."

Her voice rose, querulous. "Well, you said I could ask you anything, so I'm asking you this. Did you . . . Did you . . ." She tried three times to name something, but she couldn't. Then her shoulders stiffened, she turned her face away, bit her lips, and began to weep.

He stopped seeing April. Once he called simply to ask how she was and she hung up.

He had meant no harm. Only trying to live. Everyone else was just as bad. He knew cases. . . . People were bad from the

moment they were born. From the moment they reach consciousness they start hurting other people, in their efforts to live.

"Please," he cried out hoarsely. "Let me live!" Linda stirred. But
the mote was merciless. It would never go away. The time was 5:35.
Maybe he would die soon. He felt limp enough to die.

Maybe he had never loved anyone enough, and that was his
sin. Not his parents, not Linda, not Jack or Sandy, not even April,
not even himself. He honestly didn't know. What was enough?
Enough for what? That was too easy a solution, too glib, saying he
had never loved anybody enough. He was an ordinary man. He
could honestly say that he loved his wife and children as much as
the next man. It was something more, something beyond everything
he had thought of, cosmic, even. But what? What? Light seeped in
the window. It was dawn. Surprisingly fast, the room filled with
light. As it filled, the mote faded and abruptly disappeared. The air
around his bed became less dense. Alexander slept.

Benjamin Cheever

THE DOOR

OF PERCEPTION

Stanco is the Italian word, and I suppose that some of you are *Stanco a morire* (tired unto death) of hearing the smaller Cheevers talk about the Big Cheever: How he lived. (Brilliantly.) How he died. (With extreme reluctance.) How he loved men. Still, there is an incident having to do with sleeplessness and writing that haunts me. And my father is in it. The central character.

·   ·   ·

My mother's family has a summer place in Bristol, New Hampshire, cottages on a hill above a lake. I went there often as a child, and so the place, although it still exists, has its greatest presence in a distant past. Below the cottages there's a larger house of native stone. This was for Gram, the patriarch, my mother's father, a doctor notorious for his knowledge of poison gases. On the path that leads down to the stone house, there is a building, not much larger than a toolshed, but dignified with the name of laboratory. This was Gram's laboratory.

When I was very young, probably six, or five, I prided myself on waking before dawn, and wandering outside naked or in my drawers. There was a power in the steel morning light, and also a sense of righteousness in taking from the day before the grown-ups had even given it a name.

I was prowling that hill one morning when the door to the laboratory opened, and my father appeared. Behind him, I could see the dark interior of the building. I seem to recall that he was wreathed in smoke. Probably cigarette smoke. (In those days a man could smoke three packages of cigarettes a day and still plan to live forever.) Clearly my father had been awake for some time, alive and puissant, a mortal threat to my childish kingdom.

I don't recall now if we exchanged words or even if he saw me, but the incident seems to have made a lasting impression. There was something going on in that building in the dark, something I wanted to get at. When I was twenty-three, and nursing a massive case of writer's block, I went up to Bristol once, alone with my young wife, and in the afternoon I walked down to that laboratory lugging an enormous electric typewriter. Remember those machines? This one came in its own metal suitcase. Turn it on and the engine would thrum like the diesels of an ocean liner.

I had spent money I didn't have for that typewriter. I was a highly superstitious youth, and used to think that the way to write was to get the best equipment, position oneself sort of like a lightning rod, and wait for a storm to blow up. I knew that lightning had hit the laboratory before. At least once.

But no storm struck that afternoon. There was not a cloud in the sky. I sat there alone and bored, listening to my outrageous typewriter. Looking back on the scene, it seems as if my father must have been dead at the time, but he was not dead. He was only in New York. Which is not the same thing. No matter what they say in the Moral Majority. Still, I was sad. I didn't weep, because I am male and have been trained not to, but I didn't feel good either. After a time my wife came to fetch me; I pretended to be disappointed.

"What happened?"

"Nothing. *Niente.*"

I know now that I had it all wrong. Wrong typewriter, wrong attitude. Wrong writer.

I also know that there was something I could have done. Two things actually. I should have stayed all day. I should have stayed until it hurt. And I should have started early. Very early. I should have started in the dark.

Because I really believe there is a sweet spot in the day, a time when a writer is much more apt to meet his muse. There's a door we pass through, between sleep and wakefulness, between life and death, and it is my contention that the closer you are to that threshold when you sit down at the typewriter, the better your chances.

No, I'm not kidding. Look at Anthony Trollope. He produced forty-seven novels, most of them while holding a full-time job. "It was my practice to be at my table every morning at 5:30 A.M.; and it was also my practice to allow myself no mercy," he wrote in his autobiography. "By beginning at that hour I could complete my literary work before I dressed for breakfast."

He cranked out 250 words per quarter hour. With or without inspiration. "There are those who would be ashamed to subject themselves to such a taskmaster, and who think that the man who works with his imagination should allow himself to wait till inspiration moves him. When I have heard such doctrine preached, I have hardly been able to repress my scorn. To me it would not be more absurd if the shoemaker were to wait for inspiration, or the tallow-chandler for the divine moment of melting," he wrote.

The observation that one does better with discipline than without discipline is not news now, nor was it news when Trollope's autobiography was published in 1883. What interests me is his timing. He worked with his back against the door of night. Discipline is a slippery fellow, easy to swear by but hard to get your hooks into. Besides which, you can grab for discipline and get his twin brother, self-righteousness. You'll know because of the side effects. You'll have discipline, but also you'll begin to have trouble with your digestion, the breath will go sour, the skin gray, you'll lose all your friends. If you are married, your spouse will take a lover. But what about timing? Timing doesn't have side effects. Everybody's capable of timing. All you have to do is when you set the alarm, you set it for early.

Trollope's mother, Frances, didn't complete her first book until she was fifty. When she set down her pen at the age of seventy-six, she'd produced 114 volumes. How'd she do it? Frances got up at four.

I don't know if she had inspiration, but she most decidedly did not enjoy peace of mind. The unfortunate woman was supporting her entire family, including her bankrupt and invalid husband. Two of her children were dying of consumption. Trollope reports that "The doctor's vials and the ink-bottle held equal places in my mother's rooms. I have written many novels under many circumstances: but I doubt much whether I could write one when my whole heart was by the bedside of a dying son." But then Anthony didn't get to work until after five. I also think that he was mistaken about death. Death is another door. The closer you are to it, the sharper the prose. But back to night.

The Trollopes aren't the only ones with a tropism for darkness. Churchill preferred to dictate at night, so late at night that it was often morning. My sainted father said he wrote to "celebrate a world that lies spread out around us like a bewildering and stupendous dream." When's the best time of day to remember your dreams?

Dr. Jekyll and Mr. Hyde is supposed to have come to Robert Louis Stevenson in his sleep. All he did while awake was transcribe.

The evidence is not just in biography, but in the writing itself. Take the opening of Daphne du Maurier's *Rebecca*. "Last night I dreamt I went to Manderley again." Children's books are marketed as bedtime stories. The sleep connection is often obviously stated. "Alice was beginning to get very tired . . ."

Writer and reader both are listening to what Dylan Thomas calls "the distant speaking of the voices I sometimes hear a moment before sleep."

Ordinary men and women stay away from that threshold. They sleep, but not to dream. They'd rather not consider death until it's got them by the throat. Then, if they have a minute, they write beautifully. Fatal circumstances can transfigure the most unlikely clay.

Ulysses S. Grant was not an artist, or even a cultured man.

Asked about music he once said that he only recognized two songs. One was "Yankee Doodle," the other was not. And yet he produced, in his memoirs, a wonderfully frank and immediate book. It sold twice as well as Mark Twain's *Huckleberry Finn*. How'd he do it? Easy. Cancer of the mouth and throat. The man was dying. He had his back against the door.

I guess we all do. *Tutti dobbiamo morire*. (We all have to die.) I worked once at Boston State Mental Hospital. There was a young man on our locked ward who liked to sneak up behind the staff, when we all sat down for coffee and a smoke. "Some day you'll all be dead," he'd shout. He was right. Daylight is a miracle. Life a minute. Blink and it's over. And yet there's something more. Something inexplicable. Something fine that lies beyond the easy grasp of thought. Or even language. Approach it and even the finest mind must feel like an Englishman who has studied German in order to converse in French. Wake up and it's gone.

There's always the suspicion that what's gained in waking is not worth what has been lost. And if this is true of waking, might it not also be true of birth. "Our birth is but a sleep and a forgetting," wrote Wordsworth in his "**Ode: Intimations of Immortality from Recollections of Early Childhood**":

> *The soul that rises with us, our life's star,*
> *Hath had elsewhere its setting,*
> > *And cometh from afar:*
> *Not in entire forgetfulness,*
> *And not in utter nakedness,*

But trailing clouds of glory do we come
From God, who is our home.

Writers can go back. Wordsworth did. My father had. Writers grow to be men. They even grow to be old men. But still they work at night, near darkness. Each day remains a birth and a forgetting.

Return to that distant morning on the granite hill. Glimpse below us the large and silent lake. There were cabins on the other shore. Mountains beyond the cabins. In the evenings, the laughter of children would come to us across the water, loud and yet indistinct. There are no voices now. Dawn is just breaking. A door opens. I am startled. My father appears. Behind him I can make out the dark interior of the laboratory. The picture is vivid, and yet unclear, something seen through cheesecloth. Was his face still swollen with sleep? Was there a lighted cigarette in his hand? Had he been smoking? Or were those clouds of glory?

Michael Brownstein

THE ART

OF DIPLOMACY

I had a job once, very hectic and intense, as a press officer at the United Nations. Every weekday morning in the autumn of 1990, during Iraq's occupation of Kuwait, I rode uptown on the First Avenue bus, surrounded by fresh faces, scented bodies, crisp clothes. Then I entered another world. The long, high lobby of the Secretariat glinted with cobalt blue and silver reflections coming off the East River through a wall of windows. Gray suits moved along bright hallways. At the helm of this mythic infrastructure of some 7,000 people sailed the Secretary-General, Javier Pérez de Cuéllar.

Now and then I would see him, buffeted along by his entourage of bureaucrats. But my realm was a more circumscribed one.

Out of twenty press officers, who together were responsible for producing an official published account of everything that took place at the UN, I sat in a team of six at a big table down three steps from the speaker's platform in the General Assembly. Flanking the podium were two larger-than-life red velvet wing chairs, empty except when visiting heads of state came to the Assembly. Whoever spoke from that podium addressed a soaring hall filled with delegates of the 157 member states.

Our job as press officers for the Organization—as it preferred to be called—involved striking a balance between actual news and a delegation's desire to be heard. Writing extemporaneously while each delegate spoke, we saved only the highlights of speeches, presenting them clearly and succinctly. This meant that 75 percent of every speech, full of rhetoric and obfuscation, ended up in the trash.

Teenage runners appeared beside you at regular intervals, taking the text you were furiously scribbling out of your hands and carrying it upstairs to the editors. Then, while you were handed a copy of the next speech—perhaps the chief delegate of Hungary holding forth on the Law of the Sea—the soundless phone on the table might light up. This would be an editor challenging what you'd written fifteen minutes earlier, when you'd been relating Brunei's position on apartheid in South Africa. Meanwhile, the delegate from Hungary was saying something quite different from what appeared in the text you'd only begun to examine, so that while explaining in a heated whisper what Brunei had really meant, you were listening with your free ear over one headphone to a simultaneous English translation of Hungary's speech. We often dealt with unannounced changes in schedule, obscure references, unclear locutions, and challenges to our veracity by various delegates.

There were morning and afternoon sessions during the Plenary. Our account of each session had to be available shortly after it ended. A frenetic, tense atmosphere prevailed on the job, broken by tedious stretches during which nothing happened while two countries haggled offstage over the wording of some resolution. But when a session got going, delegates would hold forth on any of dozens of topics. Split-second decisions had to be made. At first, this was agonizing for me and the three other temporary employees at the table. It took us weeks to get a grip on the code—on the way one was expected to write—and to trust ourselves enough to ignore whole chunks of verbiage. But the two veterans who carried the table, Sandra and Denise, were able to coast along. They had memorized hundreds of stock phrases and abbreviations and could anticipate what the delegates were going to say, knowing the issues addressed, the organizations cited, even before they were mentioned.

For more than two months, I stumbled home at the end of the day utterly exhausted but wired, wide awake. I would take a hot bath, maybe drink a beer, smoke some pot—anything to relax, to get to a place where I might fall asleep. Sometimes I'd go out to a movie or have dinner with friends. But usually, no matter what I did, midnight, 1 A.M., 2 A.M. would pass, and I couldn't get to sleep. I meditated, counted sheep, drank herb tea, swallowed capsules of valerian. I avoided sleeping pills, which would have left me groggy the next day. Because I knew that by 8:20 the next morning I had to be shaved, showered, suited, and out of my apartment. Inevitably, around 4 A.M., I would finally slide off into unconsciousness. At 7:30, the alarm rang.

I caught up on lost sleep during weekends, congratulating myself on my stamina and on the big paychecks I put in the bank. By the middle of October I was dragging through work half awake. I dreamed of Thanksgiving, and especially of the Friday before Christmas when the Plenary ended.

Afternoon sessions became the real test. By then I'd shot my energy for the day and longed to close my eyes. I fantasized about the leather sofas in the basement hallways, scheming of ways to get down to one of them for a nap. "I have a headache," I'd announce, "I have to go to the infirmary." But headaches, upset stomachs, family emergencies became so common at the table that soon no one was allowed to leave without documentation of their plight. By early November, those leather sofas became objects of legend, described in hushed words of wonder among the temporary press officers.

On August 2, Saddam Hussein had invaded Kuwait. During the following months, this event and its repercussions constituted the emotional core of everything that happened in the Organization. The Security Council passed resolution after resolution condemning Iraq. Economic sanctions were accompanied by a military buildup which by November 1 involved the presence of 210,000 American troops in Saudi Arabia. President Bush dismissed the possibility of a negotiated settlement, demanding Iraq's unconditional capitulation in tones that grew more and more warlike.

I remembered George Bush well. In September, virtually my first assignment as a press officer had been to cover his speech to the General Assembly. Not thirty feet from where I sat he'd waited in one of the ceremonial wing chairs, his knee jiggling impatiently, while the UN's Chief of Protocol had introduced him. Confident, fit, full of suppressed energy, he'd fairly bounded to the dais. At that same moment, Secretary of State James Baker had materialized from a doorway and strode across the Assembly floor to take a seat with the United States delegation. Both men had radiated a sleek, superior arrogance. As Bush spoke, right hand raised, palm open, I had felt myself in the presence of a glib representative of the master race.

Now, during the tension-filled weeks in November, an anonymous xerox collage circulated among the press officers. In it, a stern-faced George Bush stood beside a microphone, his right hand raised, palm open. Out of his mouth came the words: "We SHELL not EXXONerate Saddam Hussein for his actions. We will MOBI-Lize to meet this threat to our vital interests in the Persian GULF until an AMOCOble solution is reached." The oil companies' names appeared in the form of their corporate logos.

And yet, according to every delegate on the floor of the General Assembly, with the exception of Iraq and, to a lesser extent, Cuba and Yemen, oil had nothing to do with the UN's response to Saddam. One speaker after another declared that the invasion of Kuwait, alone in the history of international disputes, amounted to an act of unrelieved aggression without any semblance of justification. Justification there may not have been, but the hypocrisy of such a position must have been obvious to everyone. Recent history was filled with similar though nominally disguised power grabs by nation-states around the globe. The only distinguishing factor now was the international community's united condemnation. I wondered where the community's response had been in Tibet, in East Timor, or among the Kurds.

My colleagues at the table, however, seemed to have no problem with the speeches that came their way. Everyone was sarcastic, knowing, politically savvy. All in a day's work. Water off a duck's back. What was my problem? Was I being naive?

Maybe if I'd been rested I could have fallen into place along with the other awake, aware people around me in the Organization. But as time went on, I couldn't help thinking that I alone was alert, I alone saw clearly. Everyone else, eyes wide open, was sleepwalking.

Late at night outside my apartment on East Fifteenth Street the silence grew pronounced, punctuated only by isolated echoes and car horns. Inside, the refrigerator hummed. The clock ticked. As November came to an end, I would lie awake later and later, all semblance of trying to sleep abandoned, and think about the General Assembly. Its perfect hollowness took my breath away. If Bush went to war, Iraqi civilians would die by the thousands. The Iraqi people would become sacrificial victims. Already, Iraq's delegation—four men, one woman—had been ostracized for months. But at the end of November, they suddenly became pariahs.

This happened on November 29, when they walked out in protest after the news spread to the Assembly that the Security Council (which met elsewhere) had just passed Resolution 678. For the first time, this authorized the Organization to use "all necessary means" to restore peace and security to the area. In other words, Bush was saying that he now had the United Nations behind him in the threat to use force. Forget about waiting for economic sanctions to work, forget about making every effort to talk first. I remembered that knee of his, jiggling impatiently.

The five Iraqis were now scapegoats, outcasts. Their eyes glittering, defiant smiles on their faces, their bodies rigid, they left the session hurriedly in a pod of anxious movement that traveled down the aisles of the Assembly hall and out one of the doors. They moved along through a force field of condemnation. Everything around them stopped. Hundreds of people stared in silence at these four men, this one woman. I became dizzy, then nauseated. I ran to the bathroom and vomited. And when I returned home, for the second night in a row I got no sleep.

"All necessary means . . ." I didn't even try taking a hot bath or drinking a beer. Instead, I sat in a chair by the window, looking down onto Fifteenth Street. Then I stared straight ahead at the wall above my bed. In the darkness, the wall squirmed like it was alive. I decided I'd had enough of this job. I would finish out the week and be gone.

At eight o'clock the next morning I made a big pot of espresso and drank it in gulps, feeling superprescient. I didn't shave or shower—what was the point? Because I'd just then decided that the end of the week was too far in the future. I would quit work that very day.

As I rode uptown on the First Avenue bus, it seemed that everyone around me was sleepstanding and sleepsitting. I alone remained awake to what was really going on: People staring into the middle distance of their lives as they wandered the sidewalks, as they waited for buses and rode uptown. Zoned out on some inner narrative or other. Staring down at their shoes. Going to work every day. . . . That's when the nature of the energy on the bus revealed itself to me: capitulation. The bowed head. That's what Bush wanted from Iraq. To bring Saddam down onto one knee. To shame him. For that he would go to war. For a game—a contest—between boys.

· · ·

Several delegates spoke that morning in support of Resolution 678. Then the chief delegate of Cuba, Ricardo Alarcón de Quesada, rose from his seat at the back of the hall. Alarcón was a cripple, one of his legs was paralyzed. With painful slowness, he made his way to the speaker's platform. He took forever to reach the podium. I had seen Alarcón before, of course, and he fascinated me. Why had Castro chosen an old man who could barely walk to represent Cuba at the UN? What message was he sending? That in this organization commandeered by America, Cuba remained helpless but unbowed?

The old man, silver-haired and ponderous, made a perfect victim. When he finally began to speak, he predictably denounced American imperialism throughout the world. Then he said the United States would go to war in the Gulf because Kuwaiti princes owned a far bigger share of American banks than anyone admitted. He finished by asking for an Arab solution to the conflict, rather than the use of force. He shuffled his papers—in less than five minutes he was through—and began the long negotiation of the steps and the aisles, which would take him another five minutes at least.

Meanwhile I knew already that, once the other scheduled speakers had their say, the United States would reply to Alarcón's speech. This opportunity for any member to respond to what another member said was called *right of reply*. There were two or three at the end of each session, sometimes more. The air was electric during those weeks leading up to the war. This latest resolution had upped the ante for a military solution to the conflict. The whole world was watching. Of course the United States would reply.

Ninety minutes later, the chief U.S. delegate, Thomas Pickering, got his chance. He stood among his delegation halfway up one aisle, an extremely tall man, vigorous and forceful. He lost no time in making his point.

"Mr. President," he said. "It is indeed unfortunate, but not unexpected, that I must exercise my right of reply to the statements made today by the representative of Cuba. The contentious language and distortions put forth by the delegate from Cuba fool no one. Cuba seeks to deflect attention from the failures and shortcomings of its own repressive revolution which, after thirty years of dictatorship, have left the Cuban people still yearning for the basic human rights and freedoms now enjoyed by the vast majority of their Latin robots. . . . Thank you, Mr. President," he concluded, and sat down.

My stomach turned over as I sat at the table, crowded together with five other perspiring, nail-chewing, coffee-guzzling press officers.

Abruptly, Sandra turned to the rest of us with a quizzical look, the bracelets jangling on her arms. Because the morning session was so hot, she'd taken personal charge at the table and things had been running very smoothly.

"Did he really say robots?" she inquired. Everyone broke up laughing, we'd all heard the same thing. In the hall, the hundreds of delegates and their staff were shutting their briefcases, standing up to leave for the long lunch break. Sometimes we had to venture onto the floor ourselves to query speakers about their speeches, verifying details, clearing up uncertainties.

"Go up there and ask Pickering what he said," Sandra told me.

Hyperventilating, dizzy with exhaustion, I hurried up the aisle, catching hold of his elbow just as he turned to leave. He recognized me as one of the press officers. Friendly but rushed, he asked, "What is it?"

"At the end," I croaked, my voice barely audible, "did you—I mean, did you say Latin robots?"

He exploded with laughter. *"Neighbors . . . their Latin neighbors. . . .* How could you people possibly have heard *robots?"* The three men and two women around him were giggling. Then all of them moved off up the aisle and I turned to go. At my feet lay a sheet of paper. Picking it up, I read: RIGHT OF REPLY TO CUBA. It was a form letter, a right-of-reply answer sheet, on which were typed six stock negative formulas to be used when condemning Cuba. I couldn't believe my eyes. Pickering had simply read aloud whichever one he thought pertained to the situation at hand. In this case, the third paragraph. There it was in black and white: "The contentious language and distortions put forth by the delegate from Cuba fool no one . . ."

The other formulas were equally condemnatory.

Everybody was playing games, I realized: the Americans, the Cubans, the Iraqis, the delegates from repressive Third World regimes who regularly excoriated the United States, the supposedly prickly and independent Europeans who always ended up following the lead of the Great Satan.

I raced toward the back of the hall, pawing my way past throngs of departing delegates. Finally I caught up with Pickering just as he was leaving the floor. I reached out and slapped his shoulder. He wheeled around. Instantly, his eyes became mirthful. "More robots?" he asked.

"And a phony time was had by all!" I shouted at the top of my lungs. The hard look in my eyes brought him up short. He backed away from me. His left hand shot up in the air and instantly three men in steel-gray suits surrounded me. They pulled me back into the hall as he made his escape.

"Why don't you change your ways before it's too late?" I continued shouting. "Listen to me, Pickering, you can dialogue with Saddam if you want to, everybody knows that. Just cut a deal and there won't be any war!"

I was right, of course. I was convinced of the truth of my words, certain they would alter the course of history. And they did. In one bold stroke I mastered the art of diplomacy. Because there was no invasion of Kuwait, there was no buildup of American troops in Saudi Arabia, there was no war. At least, I myself don't remember one. Do you?

Mark Richard

CHARMING 1 BR, FR. DR. WNDWS, QUIET, SAFE. FEE

Insomnia is easy.

When you get insomnia, you open the French door windows to the apartment you can't afford and sight down on the riffraff on the corner selling drugs with your loaded .38 Smith & Wesson Airweight, and squeeze on the trigger just enough to scare yourself; or you call the pay phone they are all standing around and make vicious remarks from your dark perch about their appearance; their clothes, their grooming, the way they grab their dicks; or you can fill condoms with water and sling them at the drunken gangster patrons of the pasta place diagonally under your apartment as they come out screaming, screaming some boisterous gangster rubbish, knotting up their ties in cheap gangster fashion, shadowboxing the terrified immaculate Chilean parking garage attendant next door to the gangster pasta place, the way they straighten his tie too tight lifting him from his green plastic chair, sending him off to fetch their Buick, two of the drunken gangster guys wrapping themselves in the long garage door chains, breathing to break them across their chests, chewing the links in their mouths, growling gladiators chained to fight bears while a third gangster turns on the garden hose the immaculate Chilean uses to wash the piss and vomit of previous gangster patrons off the sidewalk, the third gangster spraying passing cars with the hose, daring them to stop, crouching between parked cars he hoses down a passing couple, unarmed people who dare not turn and face six or seven drunken gangster guys, and then, in insomnia, you, you lean out naked and swing the water-gorged reservoir-tipped and tied-off latex condom a couple of times out the French door windows of the apartment you cannot afford, you swing the condom back and forth, building trajectory, then you let it go and peek while it sails upward spastically like a happy fat girl's breasts bubbling in the top of her cheap low-cut cotton summer dress as she laughs bouncing high-heeled down subway steps, you marveling at the wonderful dynamics of these dualities, the elasticity, the abundance, the constraint, this vulgar water-stretched latex mockery with its reservoir-tip nipple sails high over the sidewalk then begins its descent, and it doesn't strike the gangster with the hose, or the two playing Spartacus in the chains, it doesn't actually strike any of the gangsters, it explodes on the sidewalk sopping the Big Man Gangster's shoes and cuffs, sending the gangsters suddenly looking up, looking up, looking up not seeing anything up there, dropping the hose, shrugging off the chains, shouting at each other, the Big Man Gangster calm but seething, one of the little gangsters mopping the Big Man's shoes with a handkerchief, Sorry sorry sorry boss, the Big Man having tolerated the drunken child's play now finds himself surrounded by IDIOTS, pistol-packing IDIOTS supposed to dive in front of the don to take slugs can't even take a water balloon, and one of the gangster brutes, to do SOMETHING, examines the water on the pavement, looking for clues to the guy we're going to get for doing this, Big Don, Boss, yeah, he taps around in the water pattern on the sidewalk SENS-ING the water pattern, something to tell him, because he's a micro-second from pulling out his B.O. Plenty TEC nine on the creep, and then he sees with heartbroken familiarity the broken condom and before he even thinks, he's saying Lookit, a rubba! and the Big Man kicks the scrap of latex out into the street, his beautiful shoe barely missing the face of the man squatting by him, and the Chilean brings down the Buick, they shove him around and point to the roof, shove him around some more, throw a clutch of bills at his green plastic chair, Chinese-fire-drill themselves into the humped Buick sitting low with cheap-suited muscle, peeling out fast at first, then slowing, passing, one last look out the car windows at the rooftops along the street, their eyes looking high above you, you in your naked insomnia in your dark, wide-flung French door windows to the apartment you will never in your life be able to afford, you grab your dick at them, you give them the finger, you spit at them, and when they are gone, you sit and wait on the corner of the bed, playing with the pistol, waiting for the trash truck, the bakery, a robbery, the dawn.

Mary Morris

Animal Rescue

A few days after the big snowstorm, we noticed the cat in the tree.
Actually we didn't notice it right away because it was a gray-and-
white cat in a gray tree covered with snow. It looked like a lump,
just some knot on the lower branches of the great red oak that
extends itself not far from our bedroom window. Mostly at night I
heard the crying. I'd wake before dawn to whimpering sounds and
listen for a while, then wake Joel. "Do you hear that?" I'd ask as I
shook him gently, and for a moment he'd listen too, then say, "It's
just the wind, Sandy; now go back to sleep."

I've never lived so close to a tree before. Never been able to hear the breeze rustling through the leaves like someone shuffling cards. I've lived in crammed rooms over air shafts on residential streets and in brownstones where they collect garbage at 5 A.M. But last year we bought this house and the oak tree was one of the reasons why. We could stand under its branches on a hot summer day and feel its shadows and cool breeze and it seemed to the both of us as if nothing bad could ever happen under its limbs. But since we moved in, I've been amazed at how loud a tree can be in the fall as leaves blow to the ground or how the whistling through bare branches on a winter's night can keep me awake.

The first time I heard the crying we were digging out after the storm. We could barely open our back door to let the dog out or get to the garbage and Joel was worried about so much snow weight on the deck, so we had to dig out from the back as well as the front. It's not that easy, digging out, and I was tired after a few minutes, but Joel gave me one of those looks like if I can do this, you can, so I kept at it, though I made my shovelfuls smaller.

As we hurled clumps of snow down into the garden, I heard the soft whimpering. It seemed to come from somewhere nearby yet far away at the same time and at first I thought a child had been left in a basket on a doorstep on a winter's night. "Don't you hear that?" I asked him.

"No, I don't hear a thing," he said as he cut into a big drift on our little picnic table that reminded me of a Halloween ghost. I'd liked that drift and was sorry to see it go.

Early that evening we fell into bed, exhausted from shoveling out. Usually we crawl into bed early and read for an hour or two. But that night we didn't bother. I curled up against Joel with his feet and hands still very cold. His touch sent a shiver through me and I tried to warm his feet with my legs but Joel has this habit once he's falling asleep of rolling away.

I wanted to sleep pressed up against him, but he's got his shoulder—an old football injury he claims, though I can't quite picture Joel on a football field—which makes him turn on his side. I ran my fingers through his dark curly hair, which had begun to thin a little on top, but he gave a shudder, so I stopped. And then I heard it. The crying. That soft whimpering. Someone is cold or hungry. Someone wants to come inside.

I listened for a few moments. It was like someone was pleading with me, asking for I don't know what. Then I nudged Joel. He is used to me being bothered by night sounds. In the spring I have to sleep with earplugs because of the crows. For weeks at a time a flock will come at five or so and squack in the branches of the tree. I never imagined birds could make such a racket. And then in the fall the squirrels hurl acorns that explode like cherry bombs on the patio below.

When I went to the doctor for my checkup, he asked if anything was bothering me and I said yes, the crows. He looked at me oddly but I explained that they land in flocks in the tree and caw at dawn. It has crossed my mind to get a BB gun and shoot them down. But mainly it is the little sounds just before dawn that bother me—a back door opening, a cat in heat, someone turning on a hose. It is a lonely time to be up, not a time I like very much, because there's no one to talk to and because it really is the darkest before dawn.

. . .

I heard the whimpering for the next two or three nights. Someone is lost, I told myself. Someone has been locked out in the snow. I'd wake Joel and ask him if he didn't hear it. But he never really did. And then on the third morning when I looked outside, I saw it. Instead of a gray lump, two yellow eyes like a set of headlights stared straight into mine.

When Joel came downstairs, I stood at the window, a coffee cup in my hand. "See. I told you," I said. He looked outside, squinting in the glare of the snow. "Look, it's a cat. It's caught in the tree."

Joel pulled back the curtain and stared. Then he shook his head. "You're right. It is a cat."

"It's stuck; that's why it's crying."

"Well, if it got up, it can get down. That makes sense, doesn't it?" Joel said as he padded off to shower.

Joel is allergic to cats. And to dogs, though our dog is hypoallergenic (it has hair, not fur). His attacks come upon him slowly, not all at once. For days he'll be fine and you think he's cured. Then one night you lie down beside him and listen to him breathe. It is a deep, hollow sound as if what's lying beside me is some drowsy beast, and as soon as I hear it, I know that whenever I've taken in a stray, which I don't do very often anymore, it will have to go.

I had a cat when we first met. She was a difficult creature named Plymouth, after the rock where I found her. She was a feral, nasty thing, and one night, shortly after I'd moved to New York, Plymouth escaped and walked across several window ledges, sat down on an air conditioner of some people I didn't know who were away for the summer, and screamed all night long, which I didn't really know a cat could do until that night.

I didn't have much in my life then. I had a part-time job as a music teacher and not many friends, so when the ASPCA came to take the cat away, I wept because she was all I had. At that time Joel was just a friend. We'd met in a coffeehouse on the Upper West Side and we were both in graduate school. I was studying music because I wanted to be a choral director and he was doing his doctorate in mathematics and I called him sobbing and he said he'd go with me to the ASPCA and convince them that the cat was not really dangerous, just frightened, and that there is a world of difference between those two things, though as I've grown older I've come to wonder if this is so.

The vet at the ASPCA gave Plymouth a tranquilizer shot and asked me if I wanted something as well, and so, since I thought he meant Valium, I said yes. He left the room and came back with a little card that had some printing on it and told me to recite it whenever I could. It read: "I'm not good, I've been bad, show me the way so I won't be sad. Life is good. Life's a gift. Keep the faith. You'll get a lift." Joel and I both laughed over it on the bus home, but I still have the card.

When Joel and I began seeing one another, I had to go to his place because if he stayed in my apartment with Plymouth he'd get that hollow, wheezing sound in his chest and have to go home in the middle of the night. And then when we moved in together, the cat just had to go, so I found her a home with my sister's old boyfriend, though they were together at the time, who lives on Long Island.

Now, that morning when Joel came down ready for work (he teaches math at a community college, though his job seems to be always on the line), he wore the nice green sweater I gave him for Christmas and he had his briefcase and a serious look on his face. "I hope you're not going to get all worked up over that cat."

"Oh, no," I said, "I just want to get it down from the tree."

All day at work as we were rehearsing the Rite of Spring festival for the March assembly at the school where I teach, I thought of the cat, sitting in the tree. I wondered what made it go up there in the first place. Was it chasing a bird or was it being chased by one of the neighborhood dogs? Was it hunger or fear that drove it to where it now sat?

As I watched my third graders, faces turned skyward, mouths open in song, I was hoping that the cat would come down by itself, but when I got home that afternoon, it was still there. Only it had turned around again, so now I could see its backside once more, and not its yellow eyes, staring down at me. It was quiet and motionless and I could see how, for so many days, we hadn't noticed it was there.

My neighbor Bruce Weinstein was outside pinning up his wisteria that had tumbled off a trellis in the blizzard, and he gave me a big smile when he saw me come out on the deck. I called out to him, "Bruce, do you see the cat in the tree?"

He looked up at me and stared. Then he gazed at the tree. "No." He shook his head. "I don't see it."

"Look again," I told him. The oak tree sits between our two houses, which are so close together that the Weinsteins water our roses with their sprinkler and we smell their barbecue in our bedroom. We are both responsible for the care and pruning of the oak tree, which means that when the Arbor Barber comes once in the spring and once in the fall, we split the cost.

I'd have to admit that our property values would go down if anything happened to that tree. Although I never really think about property values, because it is the tree that has taught me about the seasons, about time passing, which I never thought about much when I was younger and single and living in Manhattan. I can see how a person living in the city can go through his whole life and hardly think about whether it is summer or winter, except that he has to change clothes.

Now Bruce's eyes scanned the branches, and then at last he nodded with a look of recognition. "You're right." His hand shot up like an arrow. "It's there."

"Don't you hear it, crying in the night?"

"Oh," Bruce said, nodding thoughtfully, "so that's what's been making that noise. I heard it all last night. Poor thing."

Bruce used to have his wind chimes hanging from the trellis below our bedroom window. When we bought this house I said to Joel, those wind chimes are going to be a problem. I'd want to sit in the garden and try to read and have to listen to the high-pitched chime of Tibetan bells. And sure enough, when the wind was high I'd hear them all night long and they'd ring in my sleep like distant music until I finally asked Bruce as politely as I could if he would consider moving them to another part of his garden, but he took them down instead. He told me that the bells were his daughter's idea. They'd kept him awake too.

"Maybe we should call someone?" Bruce said, staring at the tree.

"Who?" I asked.

"I don't know. Who do you call? Animal Rescue, the ASPCA."

"All right," I called back to him, "I'll do that." After he went inside, I tried the ASPCA, but got a recorded message that said to call back Monday afternoon. Then I tried Animal Rescue, but a recording told me to phone my local precinct. When I got a police officer on the phone, I told him that a cat had been in our tree for four, maybe five days now and that it cried all night long. "Lady," the policeman said to me in a rather patronizing voice, "have you ever seen the bones of a cat in a tree?"

That night I couldn't sleep. I envisioned a frail skeleton, detached limbs, pale white ribs, resting forever in the crotch of our tree. As the cat whimpered into the night and Joel slept soundly beside me, I began to think about all kinds of things I hadn't thought about in years. About how my brother had kept me out of his tree house when he was six and how my boyfriend had cheated on me in college. I thought about phone calls that hadn't been returned, letters that had gone unanswered, household tasks that needed to be performed. In the end I got up and started to go through a pile of clutter in the living room and noticed from the perfect circle of light in our back garden that Bruce's light was on too.

Early the next morning our doorbell rang and Bruce was there. Since it was Saturday, I invited him in for coffee. "Any luck with Animal Rescue?" he asked. Bruce looked tired, but then he always seems to have those circles under his eyes and that pale washed-out skin.

Pouring him a cup, which he said he took black, I told Bruce I'd only reached answering machines except for the police, who had assured me that cats don't starve to death in trees.

"I'm sure that's true," Joel said, sipping his coffee, the newspaper splayed out before him.

"Well, it's been keeping me up, too," Bruce said with a sigh. "Melinda says I'm like the Princess and the Pea. Anyway, I had an idea." I looked at him and then at Joel, who was doing that thing he does with his eyes when he pretends to listen but he's really reading the paper. "About how to rescue the cat. I was thinking, I mean, it just made sense to me, that we could tie some ladders together. I have an extension ladder. You have a ladder, don't you?"

We looked at one another, uncertain if we owned a ladder, but then I said I was sure we did. Then Bruce got excited and the paleness left his skin. His cheeks became almost flushed. "So we could tie them together, make a kind of walkway. Put some food on the end and coax the cat down with food tied to a pole."

"I don't think it will work," Joel said.

I looked at him. "It's worth a try," I pleaded.

Late that afternoon while I heated hot chocolate, Joel and Bruce tied our ladders together with a spool of twine, wrapping them with old towels. Joel kept gazing my way, as if to ask me how I got him into this, while Bruce worked furiously to tie up the two ladders before darkness settled in. It was late when the two men hoisted the ladder into the crotch of the tree, where it settled with a bang. Bruce had smeared a trail of cat food along the towels they had tied. Melinda, Bruce's wife, called to him from their yard, "Bruce, don't you think it's time you come in?"

When Joel came in it was late, after dark, and he poured himself a tall scotch. "That was the dumbest thing I have ever done."

"Well, if the cat comes down . . ."

"If it makes you happy," he said as he settled down in front of a sports event, "that's all I want." I don't really like to watch TV, but Joel can channel-surf for hours. Sometimes it eases his nerves, and that night in bed he reached for me and I wanted him as well. I was moving into him and felt him start to rise against me, but then we both heard the crying. It wasn't a soft whimpering, but came to us louder, more insistent than before. It almost seemed as if the cat had moved closer to our window and could see us there in bed. "God," Joel said, "why doesn't it just crawl down?" I moved into him again, but by then the moment between us had passed and instead we fell asleep in each other's arms.

In the morning I woke at dawn and saw that the ladder had shifted in the night and was now resting on our fence and the cat still sat, its backside to us, impassively, starving to death outside our bedroom window.

When Joel came downstairs, I was angry, though I couldn't say exactly why or at what, but I just started screaming at him. "You don't care about anything, do you?" I shouted.

"Sandy, what are you talking about?"

"You kill whatever comes around you."

"What have I killed lately? . . . Look, if it's about having a child again . . ."

"It's about everything," I shouted in a way I rarely do.

I wondered if the neighbors could hear and realized I didn't care if they could. "It's about how you only live for yourself and I don't believe we can live this way in the world . . ."

But Joel was already walking away from me, as he did the one or two times a year when I screamed at him in this way. He assumed there wasn't anything he could do about it and deep down I suppose I did believe that nothing mattered to him that much, and if that was so, where did it leave me?

As I watched him walking away, I dialed 911. I knew I shouldn't do this, but I didn't think I could take another night of that cat crying in the tree, so I called 911 and a woman answered. "What is the nature of the emergency?" she said.

I replied, "This is not an emergency, I mean it's not a life-and-death type of emergency exactly, not for a person, that is, but I need advice . . ."

The woman gave a long, very palpable pause. "It's about a cat. There's a cat in our tree. It's been there for at least five days and it cries all night and it's starving to death."

There was silence. Then the 911 operator said, "How long did you say it's been in the tree?"

"Five, maybe six days."

"Oh, God, that's horrible," the dispatcher said. "Look, it's Monday morning and we aren't too busy. I've got some rookies in the neighborhood. I'll send them over. Do you think it will need a home?"

"I beg your pardon?" I said.

"A home. I've already got three cats, but here's my number if it needs a home."

A short while later a police rescue van arrived with six rookies and an officer who said they would practice a rescue training mission in our oak tree. For a long time they assessed the situation, standing on the deck as neighbors came, peering out to see what the trouble was. Even the office workers at the chiropractor's a few doors down stopped what they were doing and went to the window. Then the policemen went back to their van, tramped through my house dragging climbing belts and pulleys and a giant butterfly net in which they intended to capture the cat. They belted up, tossed pulleys over lower branches.

When they threw the first rope up, the cat started to climb. By the time the first policeman was scaling the trunk, the cat was moving to the middle branches. As the policemen hoisted themselves up, the cat climbed higher and higher. Soon there were six policemen dangling from the oak tree and one shaking the large net, calling, "Here, kitty, kitty," as the gray cat climbed to the highest branch it could reach on the eighty-foot tree.

That evening when Bruce came over, the cat had worked its way back to the place in the lower limbs where it had rested for the past several days and the policemen were long gone, annoyed that their mission had failed. Bruce seemed less pale, somehow more agitated and alive. He was in his running clothes with a thin line of sweat along his brow.

It was near dusk and Joel wasn't home yet and I asked Bruce if he wanted to come inside. He hesitated, but then said he would. He said he'd come to collect the ladders since it didn't seem as if the cat was going to come down the tree that way or any way. I offered him a drink and he glanced at the clock on the kitchen wall, which said after five, so he said he would have a scotch and I poured one for myself as well.

And then we both gravitated to the window, where we stood side by side, gazing out at the wintry scene before us and at the cat in the oak tree, who had its backside turned to us, starving before our eyes. We stood there, like two parents staring at their sick child through the window of intensive care.

Then Bruce looked at me, a long stare, and I stared back at him until I decided I couldn't do it anymore because it was as if we were having a staring contest. But when I looked away, Bruce reached down and put his finger under my chin, just tipping my chin up toward his face. And then he kissed me, not on the lips, but on the cheek, as if I were being absolved of something I didn't even know I needed to be absolved of. His lips grazed my skin and then he pulled away. Then we both stood, looking at the floor. After a few minutes, he said, "I'll get my ladders another time," and he left.

Joel and I were not speaking that night when we went to bed and I curled into a tight, tense ball that I stayed in until dawn, when I woke with a kink in my back and neck. Then I stretched and went downstairs to make coffee. When the coffee had brewed, I poured myself a cup and sat down in the blue chair by the window and gazed out at the oak tree.

It took a while for my eyes to focus in the pale light before dawn, but soon they did and I could see that the cat was gone. I stared for a long time, scanning the higher and lower branches of the tree. When Joel woke, we stared together. At the place where the cat had been was the impression of its body in the snow.

It has been weeks since the cat's been gone, but I still wake in that hour just before the sky turns pale, thinking I hear its cry. I scan the tree, searching for that small, gray ball, though I finally seem to be getting over that. But now spring has returned and so have the crows.

Robert Antoni

DEMONS

IN THE HOUSE

Shortly before leaving Miami for Trinidad, where I would spend the five days of my Thanksgiving holiday, I got a call from Elyse Cheney about this piece for *The Literary Insomniac*. I told her I was a chronic sufferer, that my inability to sleep had a lot to do with my writing process, and that I'd written two long novels about characters who tell their stories because they can't fall asleep. She decided I fit the bill, and I made a mental note that, while in Trinidad, I'd try to come up with an idea for a story. On the plane down I even tried to remember the insomnia episode from *One Hundred Years of Solitude*, thinking maybe I'd do a parody of García Márquez. Peculiarly enough, the only thing that came to mind was all that nonsense about the "story of the capon," which the sleepless inhabitants of Macondo tell to one another again and again through the night—and which I'd never really understood anyway.

In the Caribbean, of course, we have no Thanksgiving; neither do we have Native Americans left to be thankful to. The American side of me always felt a little embarrassed by this holiday, a little angry, and—still sitting on my plane headed south—it occurred to me that perhaps I was making my getaway from those same stuffed-up Butterballs. Flights into Port of Spain arrive late in the evening; departures leave at dawn. Because of this, I've always followed the same ritual for my returns to the island. I spend my first and last nights with Uncle Reggie and Aunt Zel in Arima, a ten-minute drive from the Piarco airport. For years I've called unannounced. Uncle Reggie was always more than happy to come collect me. He was happy this time again, though he informed me that he and Zel no longer lived in Arima; several months before, they'd moved to a house in San Fernando.

I'd be the first to sleep in their guest room. It had its own veranda, looking across Circular Road to the chicken farm. My Uncle Reggie owns several such farms in Trinidad, and is best known on the island as the "Chicken King." The San Fernando farm is the largest, and the reason for their move, Uncle Reggie told me, was to live closer by. When we got to the house I made a few phone calls, arrangements to be picked up first thing in the morning. Zel started pulling things out of the fridge (they'd already eaten their dinner), curry chicken and chokha and bus'up shot. I sat down to eat—all part of my first-and-last-night ritual—Zel, my East Indian aunt, knowing full well that her cooking was one of my prime reasons for returning to Trinidad. We had a few drinks, and I climbed the stairs to go to bed.

That night I did something I almost never do: I fell straight asleep. Then, at precisely 2:30 in the morning, I awoke from this dream. I was changed into a chicken. A man I first took for Uncle Reggie—but later realized I'd never seen before—came chasing behind me with his cutlass raised, bawling, "Stop by order of the Chief of Police of Uganda!"

Needless to say, I did not sleep again for the rest of the night. I'd have to tell Uncle Reggie about my dream, I decided; he'd find it funny enough. Later I put on the night light, and I read the only thing I could find in the room—a technical manual entitled *Your Domestic Fowl.*

I never did tell Uncle Reggie my dream. That morning I found him eating breakfast with Andy Reuben, his next-door neighbor and the overseer for the farm. I sat listening to them discuss the pros and cons of flying in "Christmas chickens" from Miami, until my ride arrived.

My time in Port of Spain had little to do with what I had not yet recognized as the "poultry" theme of this visit, with one exception. On Thursday evening I went with several friends to have dinner at Shalimar, just beside the Normandy Hotel. The restaurant was filled with foreigners, mostly Americans, which did not strike us as unusual. And it was not until after we'd had a few drinks, and the waitress brought the menu, that we realized the fare for the night was a Thanksgiving special: turkey with stuffing and pumpkin pie.

I'd actually forgotten my dream until I awoke from it again at precisely 2:30 A.M.—asleep in the guest bed of my aunt and uncle's home—on the night before my departure. Transformed again into a chicken, chased by the same fellow with his cutlass, still bawling, "Stop by order of the Chief of Police of Uganda!"

Again I could not go back to sleep. Again I made a mental note to tell my dream to my uncle. But this time I didn't even bother putting on the light to look for something to read: I lay staring up at the corrugated-tin ceiling, more than a little shaken, listening to the steady hum of the chickens on the other side of Circular Road. Finally I got up and went out to sit on the veranda, awaiting the sunrise and an explosion of roosters, literally, like all hell had broken loose.

Again I didn't tell Uncle Reggie my dream. Andy Reuben was sitting with him eating breakfast, still discussing the pros and cons of the imported "Christmas chickens." They had to get to the farm right away; it was Aunt Zel who drove me that morning to Piarco airport.

And, once again, I'd actually forgotten my dream. Until I exchanged my *Miami Herald* for the *Sunday Guardian* of the woman sitting next to me. On page 5—startling me as if the roosters had gone off in our cabin—I found a picture of the house on 83 Circular Road, with the very veranda where I spent the night:

DEMONS IN THE HOUSE

The macabre story of a man who killed his sister because he thought she was a chicken.

Story and photos by Louis B. Homer.

Nine years after Conrad Bissessar murdered his sister, Doris Thomas, a housewife from Cocoyea Village, San

Fernando, has offered a startling explanation as to why he committed the macabre act.

Camille Khan, 62, revealed that "Bissessar was demon-possessed at the time he chopped off his sister's neck and threw it into the street, then disemboweled her body with a pair of scissors because he mistook her for a chicken. The truth is, the demons were sent to possess me but they jumped on Bissessar instead," said Camille.

The murder took place in September 1986, at the family's home at 83 Circular Road, San Fernando, in the presence of Bissessar's daughter, Michelle, and Andy Reuben. When he was arrested by the police, Bissessar told them, "I am the Police Chief of Uganda and Doris Thomas died from ruptured venereal disease."

Camille Khan, a former owner of the house in which the murder took place, said she was forced to vacate and sell it because "there were demons in the house and they were put there by a lady to possess me, but they jumped on Bissessar instead."

Now she feels compelled to reveal to the public the nightmares and frightening experience she had at the house. She praised the attorneys and the trial judge who commuted the sentence to life imprisonment.

Since she moved into the house in 1977, Camille said, she was undergoing a hard time with her domestic life. "Nothing was going right. I had to endure more problems than I was capable of dealing with. Because of these problems I left for the

United States for a short while. On my return to the
house my daughter dreamt that Satan had possessed
the house and I was turned into a chicken and was
killed. That dream did not make sense to me," she
said.

In 1978, however, I began hearing strange sounds in
the house around 2:30 A.M. every morning. My whole
life was a mess. The following morning I would ask my
neighbors whether they heard the sounds, and they
would say no. The sounds I used to hear were those of
roaring beasts.

"What caused me to leave the house was an
experience I had one night around 2:30 A.M. I was
asleep in the room with my daughter when I heard
roaring sounds coming from the back stairs. I got so
frightened that I shouted aloud and my daughter got
up and put on the lights and the roaring ceased. I was
so frightened that night that I thought the devil had
come to take me away," she said.

"After that incident I contacted several relatives and
friends, including Patsy Castillo, who advised me to go
and see an exorcist to cleanse the house from the
demons. I went to Tunapuna and spoke to a woman.
She gave me four quicksilver capsules and four ten-
cents to bury in the four corners of the house. I did
what she told me to do, but later sold the house and
bought another," said Camille. [*The house on 83 Circular
Road, San Fernando, was purchased from Michelle Bissessar by
Reginald Antoni in July 1995.*]

Patsy Castillo, a friend of Camille, is a newspaper

vendor living at St. Joseph Village. She said she was told by Camille about the horrifying dream.

"Camille told me about the demons in the house long before Bissessar murdered his sister," she said. "The day the incident took place I said to myself, 'Well, yes, it still have so much evil in this world,' " said Castillo.

With tears in her eyes, Khan told the *Sunday Guardian*, "The demons were meant for me, and it was I who should have been turned into a chicken and be killed," she said. "When I heard about the killing, I went to a shop on the main road and a man came up to me and said, 'Is it a ghost I am seeing?' I asked him why, and he said, 'I hear a man kill you because he take you for a chicken.' "

Camille said she was happy that the Appeals Court had shown mercy to Bissessar and commuted the sentence of death to life imprisonment. She was happy that the attorneys, Israel Khan and Alice Yorke Soo-Hon, had come forward to assist in the matter.

Khan was at the San Fernando Assizes, defending another man who was charged for murder, when Conrad Bissessar's matter came up for a hearing. Justice Douglin, the judge, explained to Khan the nature of the offense and said that Bissessar could not find a lawyer to defend him.

"I decided to defend him under Legal Aid and brought in Alice Yorke Soo-Hon to assist," said Khan:

Alice Yorke Soo-Hon, the main advocate, and Israel Khan are members of a group called Lawyers for Jesus.

Khan said he believes in demon possession, after he
had witnessed Dr. Aeneas Wills (former judge), now a
minister of religion, cast out evil demons from a man
who was possessed.

According to Khan, "This happened in Soo-Hon's
office. At the time, the group of Lawyers for Jesus
were holding a prayer meeting when a vagrant walked
into the office, and on hearing the name of Jesus he
started to speak in strange tongues."

In preparing the defense for Bissessar, Soo-Hon came
to the conclusion that he was possessed with demons
at the time of the murder. A team of defense lawyers
researched English jurisprudence to prove insanity by
virtue of demon possession.

Khan was determined to break new ground, since he
had successfully defended a man for chopping another,
whom he thought was a "lagahoo."

The case for the defense was: "Bissessar was a kind,
considerate and loving person. There was a good
relationship between him and his sister, and strong
evidence to show that his act was entirely motiveless
and irrational. At the time he committed the act he
was insane as a result of demon possession and he was
suffering from the disease schizophrenia, which
affected his mind to such an extent that he was
laboring under a 'defective reason' and he did not
know what he was doing, or if he did, that he did not
know that it was wrong."

In reply, the state attorney submitted that, at the
time of the incident, Doris Thomas, Michelle Bissessar,

Conrad Bissessar and Andy Reuben were living together in a house on Circular Road, San Fernando. Michelle, a state's witness, testified that she saw her father go into her aunt's room with a knife in his hands and stab her.

According to her evidence, "My aunt fell to the floor. My daddy then stabbed her in the abdomen. I called the police. While I was standing in the roadway I saw my father raise a cutlass in the air. Then I saw my father lift up my aunt's head in his hand by the hair. Before coming downstairs he threw my aunt's body over the veranda.

"Then he came down with a knife and hammer in his hand. He went back to the house and went inside and came back with my aunt's head in his hand. He rested the head on the ground. He sat on my aunt and started to cut open her abdomen."

Michelle's evidence was corroborated by Andy Reuben, who also witnessed the murder. According to Khan, "Justice Koylass, the trial judge in Bissessar's case, would have nothing to do with demon possession and he prevented Bishop Mendes of the Catholic Church from giving evidence."

Defense attorneys had submitted that Bissessar had killed his sister while under a spell of demons, but this was rejected.

The state contended that "Bissessar was playing mad and he was a normal human being who brutally murdered his sister."

"The jury convicted Bissessar of murder because they

---

were afraid of him. They wanted him executed," said Khan.

When the matter came up for a hearing at the Appeals Court in November, Bissessar's death sentence was quashed and a life sentence imposed instead. Appeals Court judge Justice Mustapha Ibrahim, while delivering judgment, said it was the most brutal murder he had ever presided over during his time on the bench and as an attorney and prosecutor.

"It was clear from the evidence that Bissessar was a madman," said Ibrahim. The state's argument was "that the jury's verdict was reasonable and safe in accordance with the evidence and that, in law, demon possession is not included in the defense of insanity."

Bradford Morrow

THE NIGHT WATCH

And besides, you are dead, my dear friend.
It is not your fault, of course, but none
the less you are dead and buried.

RUDYARD KIPLING,
"The Strange Ride of Morrowbie Jukes"

The dune was shaped like a bowl. The bowl was made of sand. The bowl was broken on one side, and along the broken margin ran a slow brown river.

The bowl was deep and so, it seemed, was the river. The sky soared above you in a kind of empty euphoria, unspoiled by clouds and unblemished but for a derelict crow that circled and circled, casting its small shadow on the face of the sand at your feet. The sky was a kind of bowl, like the dune, but upside down, pure blue, and well beyond your reach, or the reach of anyone else. The sky was deep, deeper than the dune bowl, and probably deeper than the river.

You can remember a time when you did not know this place, but now it seems as if you have lived in the bottom of the bowl, on the floor of this desert basin, forever. Along its western edge where the ramparts of sand give way to the riverbank, you have watched the sun sink into the silver bar of water many times. The river has been your focus ever since you gave up on the idea of trying to crawl up the steep soft sand walls of the bowl to escape. Others who dwell here call the river Sutlej, but you have no name for it. And yet, the river, your hope, is so wide that the far shore is not visible from where you stand.

You have, some nights, stood back from yourself and watched as you lingered at the bank of the river Sutlej, studying its shoals and shallows, its hassocks and hourglass faggots of brittle river grasses, looking for a way to escape. You have seen, past your shoulder, the curious boat moored out in the river tide, and have heard the faint report of the rifle when the faceless guard fired a round to warn you not to attempt to swim to freedom. You have watched yourself back away from the bank, and have rejoined yourself to wonder once more how you could ever have come to such an impasse. But then, as often as not, the sun, having just set, dawned again behind you, over the verge of sand, the long crown of slope. And as the bowl filled one more time with light you despaired of ever being free from this place.

You twisted yourself in your sheets, soaked through with your sweat, and opened your eyes into the pitch dark. You may have moaned or called out the name of your mother or father, but neither heard, since no one came. You turned your pillow and settled in at another angle to try once more to sleep. The dune bowl rose into view again.

The dune, the sky, the river. Here was where you returned so many times, to the same dreamlike-but-not-dream landscape, because this happened to be the residence of your night thoughts. Sometimes you arrived, like the character from the story your mother read to you, on the back of a horse that had gone mad and in its lunacy carried you across the blind nocturnal wilderness until the two of you suddenly caught flight and plunged headlong into the unseen basin. Other times no horse conveyed you: you were already there. Often you were alone in the horseshoe-shaped crater of sand, which was the length and breadth of a small drained lake, with sheer embankments rising behind you and at both your sides, and the slow river before you. Now and then you were joined by others, hideous tenants who had no more hope of escaping than you. They had different faces each time you visited the dune bowl, and sometimes you recognized them for friends or enemies, and other times you did not know them. Once, your mother was there with you, but didn't behave like your mother. When you ran up to her, filled with excitement and hope that she would know the way out, she looked at you with numb eyes, before turning to walk away.

How you wished you could come to think of this place as a sanctuary, an oasis or asylum. You understood from the beginning, though. You saw it for the prison it was, is, and will be tomorrow. The bowl was the loneliest place you knew. Your sporadic companions were mute, the river made no sound, nor did the sand, which was never blown by any wind. The bowl was a quiet site, and would be peaceful but for the underlying sense it manifested—the uneasy sense that violence could at any moment and without the slightest warning come to pass here. Swift, brutal, remorseless violence.

And then you woke from light into dark.

You live in a land where most are buried and the rest are waiting to be born. A paraphrase of Matthew Arnold paraphrasing Shakespeare. You live in a world where most are asleep and the rest are awake. Only the insomniac journeys in that middle kingdom where one is neither asleep nor truly awake, though sometimes he believes himself to be awake, very awake, the most awake he has ever been. It is a delusion. He is no more awake than the unborn in Matthew Arnold's poem are waiting. They are not waiting and the insomniac is not awake. The night wanderer lives in a middle realm, possessed of wondrous undesirable powers, informed by neurons sparking vivid jolts of consciousness in a brain where under more mundane circumstances of sleep they would be creating dreams for the sleeper to watch. The betweenness makes the insomniac different, though. About that there can be no question. Colors and sounds are modified in the world of the sleepless, some bright and clamorous, others muted. Thoughts run precipitous and furious, phlegmatic and often repetitive. It is the in-between of the zombie, the vampire, the comatose. But without all the glorifying mythology or medical extravagance. The insomniac is as lonely as the comatose, as alienated as the zombie. And yet he is neither pitied nor feared. He simply cannot sleep.

The dune put you in mind of a bowl. A broken bowl. Or as if you molded a pillow from sand and then tried to sleep on it but, having tossed and turned through the long night, burrowed away with your head so that what you created was a basin which perfectly fit your skull. A broken bowl, a crushed pillow of sand.

It would never be far from your thoughts that you could not escape, because this was an interminable sojourn. You knew that, however much you hoped otherwise. You knew you were held against your will down in this roofless catacomb in the desert, with steep walls of sand hove high on three curving sides and on the fourth a river, brown and sluggish with swirling surfaces which might remind you of photographs you have seen of galaxies, Catherine wheels of muddy, slimy water. It would never be far from your thoughts that the river promised a second death to the swimmer who would dare try to escape this little enclave of the half-dead, among whom you now must count yourself.

You hardly believed it possible you could have gotten into such a quandary as this. You had on your hands many hours to rehearse the details of your short life in search of whatever iniquity or transgression, what sin or outrage you had ever committed that would justify fate to deliver you here. What had you done to deserve this? Who were these fellow inmates, some women and some men and one strange child, dressed in tattery rags and stinking of hell itself? What was this almost-nightmare in which you were quite awake and aware of your wakefulness, disgusted at the hunger and thirst that gnawed at your belly and was only quelled by roasted crow and rank river water drawn in a pouch made of the entrails of your old horse, whose name you had now forgotten. So long ago had it been since you toppled into this place while riding her, back in that other time when sleep was sleep, wakefulness was wakefulness, dream was dream, rather than this irrevocable in-betweenness that robbed you, as it robbed all your awful comrades, of your humanity. Yes, you found it hard to believe you could have wound up in such a pit, a dead end. But here you were.

Again you began to speculate. The dune was shaped like a bowl, you told yourself. And if you counted all the grains of sand in that bowl would you finally be so exhausted that you would sleep? It was not for you to know. You could know only a few things when you were here, and the rest was either of no use or beyond your grasp.

Like a bowl the dune was shaped. That is what you knew. Like a broken bowl. That is, on three sides the sand rose toward the sky at a quite specific angle of sixty-five degrees. When you went mad under the midday sun that reflected off the glazed bits of sand, so many innumerable and infinitesimal mirrors, and threw yourself at the steep walls to climb up and up toward freedom, the sand failed to hold you. What irony, what horror. The harder you clawed and scrambled, the faster you slid back down to the floor of the dune valley, rewarded for your effort with a mouthful of sand, with sand in your eyes, your nose. Sand stuck to your wet body. Sand clung to your hair. Sand spangled you. Sand held you like a diffident lover.

· · ·

Was it a story read to you by your mother, or was it a memory older than yourself, that prompted this scene all those nights when you laid yourself down to sleep but could not? The dune bowl, the river, and the unfathomed breadth of sky had become such a presence in your half-waking life that such a question seemed peripheral. You knew the answer, anyway. The scene was from the story by the writer Rudyard Kipling, and the story was a memory older than yourself.

It was as it was. Which is to say, there was a dune shaped like a bowl, a bowl made of sand, a bowl broken on one side so that the walls of the bowl tapered down to a flat shore, along whose edge was a river, a great brown river.

Insomnia, you came to learn, is an echo. And knowledge is not always power. You knew these things, knew their origins, but could not break their spell over you. You understood the story, but the story understood you better than you it.

The dune was shaped like a bowl. The bowl, which was broken, was made of sand. The cold sun rose low and white over the high rim of the eastern bank above you. Or, rather, the moon. It was the moon, you could tell by the way it gazed down with contemptuous smile and plattery eyes. The day stars blinked, silver pellets, as if some primordial hunter had aimed high his gun and sprayed birdshot into the purple flesh of morning. It was the beginning of night, then, the morning of the evening. In the sandy bowl, you watched as the others began to crawl forth from their tunnels. Some went down to the bank of the river to wash the sand from their eyes. Most walked in circles around the edge of the bowl to see if there was any change in the architecture of the hollow.

Like the dead who cannot see themselves in a mirror, you no longer would cast any reflection. So you imagine. You don't want to see yourself, anyway. The sand reflects the sun with its mica faces, and though there are many they are too small to hold your image. This is a blessing. It may be the only blessing allotted you by the stubborn sand.

Grateful there are no mirrors here other than the mica faces on the polished flecks of stone, you walk, barefoot now, having traded your boots for the meat of a beast that had bit a man who waded into the shoals in the hopes of finding a way out. Here is the story. There was a dune shaped like a bowl. The bowl was made of sand. The bowl was broken on one side, and along the broken margin ran a slow brown river. A man was trapped in this bowl and longed to escape. Seeing he could not climb the walls of sand, he walked to the shore of the great river. He had not waded very far before the quicksand caught one of his feet and began to swallow him. Several others had gone out to the shallows, not to save him, but to watch him descend into the sands. That was when the snake came out unexpectedly from a stand of bulrushes, whipped across the face of the water, and attached itself to the meager shoulder of the wretch, who by then had even given up thrashing and upon whose lips formed no more piteous shrieks.

The witnesses were as quick about their business as the snake had been about his and, disregarding the peril of their gambit, plunged in after their quarry and killed it straightaway. They paraded back into the village, holding its limp body over their heads, in as high spirits as you had ever seen here. That was when hunger overcame your common sense. What would they accept in trade for this snake? you asked. You didn't give the matter a moment of thought as you unlaced your boots, which you hadn't removed for all the days and nights you had been here—more than you could remember now—for fear they would be stolen. And you ate the savory meat that same night, not pausing to consider you might be hungry again tomorrow, and not attentive to the fact that no one dared walk barefoot on the scorching noon sand of the basin. The burns you were going to suffer would be far worse than any pain presented by mere hunger. Hunger was, after all, but a craving.

Yet, something strange has happened. What was this snake and who was this dying man if not characters in a dream? You have had a nightmare, which means you have slept. However disturbing the experience of watching the snake and the man, however disagreeable the thought of trading your boots for the meat of the snake, it was a glorious indication that, for some minutes at least, you had given up one place of consciousness for another. And what was your proof that you had dreamed? The man who drowned in the quicksand, the man murdered by the snake—he had cried out for help, hadn't he? He had made sound in a silent place.

When you are older you will learn that Kipling was insomniac, too. Kipling, Kafka, Balzac. Edison was a lightbulb, seldom closing his eyes at night. Napoleon scoffed at sleep. Van Gogh with his crown of candlesticks painting into the night. Pushkin, Churchill. Baudelaire lying on a cot facing the wall in Asselineau's room, deep in debt, unable to work, wide awake. Ovid, Homer, Virgil knew about the night. Atlas never sleeping. So many just like you. Macbeth was an insomniac from guilt. After murdering Duncan in his sleep, he walks the stage in an agony. You knew his mad words by heart: "Methought I heard a voice cry 'Sleep no more! Macbeth does murther sleep' . . . sleep, that knits up the ravell'd sleave of care . . . balm of hurt minds, great nature's second course, chief nourisher in life's feast."

Kipling murdered no one, but he too was denied balm, kept from a portion of nature's second course. The history of humankind is littered with insomniacs, most of them nameless, uncelebrated sufferers laid low by want of a nightly reprieve. You were one of the nameless many. Like a soul deprived of speech, you simply walked your weary circle, crushing the earth beneath your feet until it became fine sand, and left graveside and rampart utterances to the fantastic specters of Shakespeare, who himself knew the night as well as any ghost.

The dune was shaped like a bowl. The bowl was made of sand. The bowl was broken on one side, and along the broken margin ran a slow brown river. You looked and thought, these are the elements of which the world is made. Earth, air, water, fire, sleep.

The five elements. Earth was sand, air was blue, water was a slow brown river, fire was under your feet and over your head. Sleep was an element upon which you could not walk, of which you could not breathe or drink, by which you could never be burned. Its immateriality was not, from your perspective, such a wonderful aspect of its nature. It was more mysterious than ether.

You would not mind, in other words, being burned just once by sleep.

. . .

The dune was a bowl formed of sand. Broken along the edge where
the sun set in the evening, the bowl surrounded you on three sides,
and on the fourth flowed a mighty brown river. Perched on the slow
current was a boat with crude cabin and forward bow that rose up to
a point, much like a talon. Above, an arrogant sun. The horizon was
liquid mercury.

You studied something in the palm of your hand. You had
remembered a proverb, or phrase of some kind. To see infinity in a
grain of sand, was the proverb, or phrase. You stared at the sand
fleck, intent, determined, even hopeful. Perhaps a message was en-
coded on its surface or hidden within the grain. If so, you had been
committed, for a time at least, to discovering what that message
might be. Naturally, you hoped against hope that what was to be
seen was a map, a point of departure, a sign. But though you contemplated this grain with diligence, even a certain enthusiasm, the
sand refused to act as a touchstone. There was no infinity in this
grain of sand, quite the opposite. Here was monotony; here was a
monolith. You flicked the tiny pebble over with the fingertip of your
free hand, observed its underside. The contour and striation was
perhaps somewhat different, but no side proved more instructive
than any other.

In disgust you blew the grain of sand off your palm.

When you looked up, for the first time in some time—you did
not know how long you had been contemplating the grain of sand,
for the sun seemed not to have moved in the sky at all, though your
neck ached from staring for so long at the riddle in your palm—you

saw that a man was standing nearby, studying you. He said nothing but you saw the combination of scorn and pity set in his face. You turned, hastened toward the river. What could you have said to the man? Were you to stand aside from yourself and observe your doings in the valley bowl, you would look on with pity and scorn, too.

The man, your double, followed you down to the graceful sweep of shore, at some distance. You could feel him behind you. No need for you to look over your shoulder to confirm he was trailing you. You heard nothing, of course. His footsteps in the sand were as hushed as your own, and you could not even hear the rush of your own breathing, let alone his. Because the sun cast only the smallest pool of shadow at your feet, being high in its early afternoon perch, no shadow extended forward across your path to betray his presence. Your double. You simply felt him.

The boat rode the steady, tranquil tide, anchored a hundred yards out. With hand to your forehead, like a false salute, you gazed across the face of the river and saw that the man with the rifle was gone. Perhaps he had retreated from the midday heat into the wooden cabin to nap. Perhaps he had decided to take a swim on the opposite side of the boat. In either case, he was not at his usual post.

Without thinking, you reached down to untie your laces. You needn't have bothered, as you were barefoot. What you had traded for your boots should have reminded you of the quicksand and other perils of the river, but you either didn't remember or didn't care. The reeds brushed against your bare legs and then your arms as you waded out into the shoals. The water was cooler than you might have imagined. The river mud was soft under your feet. Soon enough you found yourself afloat, and your heart began to beat hard in your chest as you swam, quietly and quickly as you could, out away from the encampment in the dune.

You kept an eye trained on the boat, but it was as if it was deserted. You waited for those back on shore to begin calling out to you, but everything remained quite still except for the subtle hint, nothing more substantial than an instinct, that he was there behind you in the water, your double, escaping with you, escaping this unbearable place. Rather than look behind, it was imperative you continue to make your way forward, only forward, and so you did, pulling yourself through the warm and cold currents of water, which out in the river proved not to be brown, but clear as crystal and silvery on the sun-struck surface.

Once you were out some fifty yards you simply floated in the current which carried you downstream. Even here you could not see the farther shore, and gazing downstream saw no beach nor anything other than the crisp curved back of the river. The boat began to disappear as you drifted southward in the flood. Your escape was happening with undreamed-of ease.

A crow cawed overhead, and you smiled to think of the wonderful homecoming feast awaiting you upon your return. You would have quite a story to tell. When you looked shoreward, you could see that the dune bowl had altogether departed from view, and you began now to swim back toward the bank. Just before you did, however, you took one parting glance at the boat. The guard was back on deck, and you noted he was waving his arms. Others appeared on deck with him, and it seemed they were struggling to weigh anchor. Surely they would not get underway in time to catch up with you. Out of the corner of your eye you did see your double, who had fallen behind farther than you thought.

As you reached the shallows, you heard gunfire. You crouched behind a sandbar, watched. The boat had sailed downstream and the rifleman had you in his sights. You noted the black stain on the river between you and the boat. When you lifted your head to make a run for the shore you were dazed by what you saw.

The encampment, just as you'd left it. The bowl of sand. The others there, wandering along the margins of the dune valley. And behind you, the boat. The boat high on the back of the river, anchored just where it always was anchored. The stain was gone. You stood on the beach, dry in the dry air.

The bowl was never colored, though sometimes you liked to think of it as pink, the dry pink of an old shell or faded rose. The river ran slow and muddy and, telling from the tranquil composition of its surface, was very deep, and you knew that even in the moonlight the color of the river was the brown of cocoa, chocolate brown like pudding but unreflective and unsweet, a death brown. In the vision, however, the river was pewter and never picked up so much as a shard of sky as it plodded along in its unchanging channel. And what of that sky? You had to admit that there were moments when it gathered to itself the most brilliant blue, deep and luxuriant blue, a blue you were not allowed to touch.

Your food was black. The water that quenched your thirst was gray. The sand world where you lived was white, a dingy white which you managed sometimes to will toward ancient pink. Your burrow, where you lay, or cocooned while failing again to sleep, was so dark that to assign it any color at all would be a mistake. It was the darkness of the valley of the shadow of death.

And your fellow prisoners were not as white as the face of the moon or winter sun, but they were blanched almost to transparency. Isn't it true, in fact, that on more than one occasion you found you could hold your hand to the horizon and see right through the skin, see where the long straight line of sand met the long equally straight line of sky?

The truth is you weren't dead. Neither were you asleep or awake. You were merely captive. Captive to the story you were being told. "The nature of the reeking village," your mother continued, "the nature of the reeking village was made plain now."

You shifted in your bed, aware that something important was about to be offered. Yes? you whispered, to fill the hollow of silence your mother left then. You looked up at her fine round face and walnut-black eyes in the small coral light of the bedroom. Around this tight sphere of light was heavy night darkness.

"The nature of the reeking village was made plain now, and all that I had known or read of the grotesque and the horrible paled before the fact just communicated by—" But you didn't really understand what came next in the story. Things about Brahmins and Bombay, Armenians and Hindi, things about catalepsy and epidemics of cholera. And so you drifted along with Gunga Dass back to the sand place, where you did understand most of what you saw, that the sand was contoured like a bowl, and was broken along its western edge. Too, you knew, and without even looking to verify its presence, that a river was there, very wide and slow. And as your mother's voice slipped away over the crest of the sand pit until it was no more than the rhythm of memory, fainter than a dying heartbeat, you began once more to enter into that sleepless median between being awake in your bed listening to your mother read you a story and being asleep in your bed, your night light shedding subtle gold along the baseboard where sometimes you could swear a mouse would run.

The next day, having made your bed and gone outside into the waking world, the bowl of sand and the mouse would be equally real to you. You could, if you chose, get on your hands and knees and peer into the small arching hole in the wall where the mouse lived. But look though you may, you could never find so much as a grain of sand in your bed.

Therefore, a question came to mind. What did you really know about the place where you went at night? For instance, who were the others there? And you, who had never been to the desert, what did you know about endless dunes and fiery sun? The cotton-wood out your window shaded you on the hottest days of summer, and the grass in the long meadow behind the old house where you lived, these gave you no insight into the sand pit. Except for the time you fell into a ditch the road department had dug in order to lay pipe, and were stuck down there unable to crawl back out for a few hours, trapped a mere ten feet down in a rich red sepulcher of wet earth until someone heard your screams and brought a rope to save you—other than that once, what did you know of traps and imprisonment?

Were it a dream, your mother would have called it a recurrent dream. As it is, you kept it private, as if it was some precious, secret toy, dangerous even to yourself who created it, and lethal to anyone else. It would not be right to say you enjoyed the long hours spent in its thrall, but you had become thoroughly used to it. The dune by the slow river was as much your home as the room in which you strived to sleep. More so, since you observed the dune with far greater care. Someday, like the writer Rudyard Kipling, you might be able to tell the story about the dune, the river, the people there. But now, you would only watch and try to remember.

The dune was shaped like a bowl. The bowl was made of sand. The bowl was broken on one side, and along the broken margin ran a slow brown river.

The bowl was deep and so, it seemed, was the river. The sky soared above you in a kind of weightless ecstasy, and it would have been a clear sky from horizon to horizon but for the cluster of clouds that had piled into a churning gray froth out across the broad face of the river.

The boat was nowhere in sight. Perhaps the boatmen, having seen the cloud bank building out to the west, had decided to weigh anchor and put into port for the night, and thereby elude the coming storm. You had never seen the river without the boat on it, and now recognized that it wasn't the same river without the presence of the boat. That is, the river seemed as incongruous without its boat as the boat would seem without the river holding it afloat on its calm face.

The clouds churned, seethed, smoldered. You looked around you and saw that the others had retreated into their tomblike pits and then you looked again toward the western sky and saw that the clouds had grown and multiplied until they'd reached across fully half the sky. There was a wild wind now, too. The sand blew in huge yellow eddies and funnels and licked its dry tongue on your cheeks.

The sand began to bury the village, and as it did the water churned and rose in toothy waves along the short coast, gnawing at the sandy shoals. The river grass blew in hanks like razors through the air. The river itself began to walk up into that air, even as the sand began to fill the river. You looked down to discover sand covered both your feet and calves and your knees were buried in a thousand thousand tiny mirrors of mica. Soon enough the sand covered you entirely and the river vanished and the sky vanished under the weight of sand. You felt the snake curl around the length of your buried body and begin to tighten, for you breathed out easily enough but when you tried to inhale there was not room inside you for the air to enter.

What is more, the air had turned to stone.

Nothing but a dream, of course. You fought to stay awake in your bed but did not succeed. There was never a moment when the air had turned to stone or the sand encircled you like a snake. It was a falsity that you were ever short of breath. The sky remained as unblemished as ever, the dune was forever shaped like a broken bowl. The river grass was gently nudged by the ponderous tide where it emerged in clumps, and certainly was not ripped by the roots and blown by any gale. And the river was running to the west of the basin, under a perfectly cloudless sky. On the river there was a small boat that had never weighed anchor and had never for a minute left its place to sail for port. Indeed, there was no port at which to call. On the deck of the boat you could see the dark figure of a man who cradled in his arms a rifle. The scene was just as it had been. It had not changed and your role in it was as before, to walk the borders of the valley and the bank of the river, sometimes thinking about climbing the dune walls, sometimes considering whether you could swim to freedom across the width of brown water.

The dune was shaped like a bowl, plain and simple. That is, the dune had features which were shared with those of any bowl. The sides of the sandy basin rose in a curve upward, just as do the sides of a bowl. A bowl holds in its hollow whatever contents are put there. And you were the contents of the dune bowl, surely you remember that much, do you not? The dune was shaped enough like a bowl to merit comparison with a bowl.

The dune will always be a bowl, even though you would want it to be shaped enough like something else so that the insomnia that comes from this going round and around about the dune, the bowl, the sand, the river, might come finally to an end. But it will not come to an end because children never want a story to end, and they never tire of hearing the same story over and over. They never tire of hearing how the dune was shaped like a bowl, how it was made of sand, how the bowl was broken on one side, and along the broken margin ran a slow brown river. And however tired you may be, a part of you will always want to hear the story about the man on the horse that went mad and carried him away across the desert to this place of sand, sky, river, sun, and of the mysterious element, the fifth, called sleep.

RECKLESS

DISREGARD

A man went to his psychiatrist and said: Doctor, I don't know what's wrong with me. I'm a tepee, I'm a wigwam, I'm a tepee, I'm a wigwam, I'm a tepee, I'm a wigwam. The psychiatrist said: Relax, you're two tents.

The noise was killing her. She wanted to murder them. She knew her life would come to an end before she bought a crossbow and arrow, before she murdered the morons throwing garbage cans. She saw her end. She'd be murdered. She thought about murder too much. Whatever you think about comes to you. It becomes you, you become it. Violence would seek her out and claim her body. It had claimed her mind. She'd be its victim. It would be her just deserts. Or unjust deserts.

She couldn't sleep. She couldn't see herself going into Paragon Sporting Goods store and asking to look at crossbows and arrows. Before she did anything, Elizabeth saw herself doing it. If she was going to walk down the stairs, she saw herself walking down the stairs. She saw herself taking the first step. She prepared herself. Her heel might catch in the hem of her pants and she'd hurtle forward and land on her head. She could decide to jump, lunge, leap, or fly over the stairs. She thought she could fly over a flight of stairs, if she tried. It looked easy. She wanted to try. She didn't want to train for years to be able to do it. That was crazy.

She wouldn't murder the morons in cold blood or in a moment of passion, whatever that was. When she murdered, it would be in self-defense. She'd be attacked. A large man or a small man would come at her. From behind. Moving quickly, she'd swing around, arms flailing. She'd gouge out his eyes or jab her fingers into his gut. She wanted to be able to sever someone's jugular vein or hit someone over the head with the baseball bat Henry kept near the door. She'd bash the aggressor to death without blinking an eye. Then she'd toss the bloody bat onto the floor and phone the precinct.

I just murdered a man with a bat. Yeah. Right. He's bleeding. But he's dead. Don't send an ambulance. He's bleeding, but he's dead. Yeah. A bat. A baseball bat.

Even her revenge fantasies were silly. They ended without conviction.

With tired eyes, she followed the band of morons. They sauntered to the park. They turned over another garbage can in a blasé way. Threw one at a car. They'd had a lot of experience throwing and overturning garbage cans. They turned over the last one casually, even gracefully, with a little wrist action. They could be tennis players or garbage collectors. In her neighborhood, the garbage collectors left as much garbage on the streets as they picked up. They threw the garbage cans all over the sidewalks. It was a display of real disgust, gutter hatred of the poor.

Elizabeth caught them doing it. On another night she couldn't sleep, she went downstairs at 6 A.M., carrying newspapers to be recycled. The garbagemen were throwing garbage and garbage cans. The street was an ordinary disaster, strewn with evidence of rampaging dogs or mad people. She wished she had her camera. But the garbagemen could argue about the photographs. They'd get lawyers, they'd interpret it their way. Her block wasn't covered in garbage, it was her point of view, how she saw things, she had a distorted view of the world, of the block. She did.

They'd say the garbage collectors couldn't have done it, because they were on their coffee break. Some hooligans must've done it, they fled before anyone saw them. Elizabeth could spend her life in court defending herself, her story. She'd present her story, and one of the garbagemen would say: That's not the way it was. He'd shake his head adamantly or sadly, as if the thought of his doing something like that was beyond him. I would never do something like that, he'd insist dramatically. Maybe he'd cry. The jury would side with the men in uniform. Elizabeth would be branded as a lunatic, an urban malcontent.

She remembered the garbagemen down at the end of the
street in their uniforms. She remembered their faces. She remembered thinking: I pay taxes to the city for them to take away garbage. It was pathetic. She watched as they flung the last cans onto
the sidewalk. She surveyed the devastation and then glared at the
men. She memorized their truck's number. She was overwhelmed by
despair. She noticed the alcoholic super down at the other end of
the block. His face was enflamed, scarlet. Sometimes his face looked
tanned and healthy, sometimes like an old shoe. She walked over to
him, he always knew everything, who was in jail, when there was
going to be a bust, who was going to jail and why. He knew she
liked the street stories. Elizabeth announced that she was going to
report the garbage collectors. He said:

What'd they look like? A tall black guy and a short Italian
guy? The regular guys are O.K. These aren't the regular
guys. The regular guys are good guys. They wouldn't do
this.

He gestured to the street. They both looked at it and Elizabeth asked: Are they rogue garbage collectors? The alcoholic super
and Elizabeth laughed in the mordant morning air. Morning is for
mourning, Elizabeth thought. Another garbage truck rolled along
and disgorged the regular guys. They were doing the other side of
the street. Elizabeth walked over to the short Italian one.

Take a look at our block. It looks worse than it did last
night. Look at the garbage everywhere, look at the cans all
over the sidewalk. How can they do this and call themselves garbage collectors?

The regular garbage collector surveyed the sidewalk. He saw the randomness, the mayhem, the sidewalk littered haphazardly with black plastic and aluminum cans. He saw the Chinese food, milk cartons, dog shit, cat-food cans, and diapers scattered contemptuously on the ground. The regular guy hurried. He raced to make things right, to turn the cans right side up. He shouted, as he ran, that he'd take care of it. He didn't want her to report them. He didn't want trouble. She didn't report all the wrong things she saw. It was depressing and time-consuming.

Elizabeth looked out the window. She watched the morons again. They were crossing Avenue A. They were screaming. A speeding cop car or an ambulance racing to save someone could hit them. They might die. They could all be murdered in the park by a fellow moron or a desperate addict. Her mother said: Where there's life, there's hope. She didn't want to die, she told Elizabeth, because there's no future in death.

The third-floor man was still in his window across the street. Even with his lights off, she saw a dark shape. It filled the window. It could've been his dog. Henry was sleeping. Strident, bizarre noises didn't wake him. A series of high-pitched yelps or squeals started. They seemed to come from someplace close. It sounded like someone was being tortured. Henry didn't move. He was a smooth stone on the bed. He didn't look alive. Elizabeth couldn't figure out if the torture noises came from human beings, dogs, or cats. People tortured their animals. They tortured their children. Children tortured animals. Everyone's a monster, given the opportunity.

The man was watching her from his window. He was pretending he wasn't. She didn't want to hide. She was covered, decent, whatever. He was probably the kind of man who made ugly noises

when he slept, when he fucked, the kind who wears sickening perfume. Maybe he knew he was a creep. Maybe all creeps know
they're condemned for life.

The young super from the building on the other side of the
street walked out the front door. Onto the street. He glanced from
east to west. He played the role of an important man expecting
someone or something. He couldn't have expected to catch the
morons. They were long gone. He shuffled in an aggravated way to
the overturned garbage cans. He saw the damage. He cursed loudly.
His arms flapped up and down, jerking out from his body. He
checked his car. It was O.K. The one next to his was dented. He
didn't react. The garbage can throwers weren't on the church steps.
The young super took his time. He was a creep.

When the young super first took over the building across the
street, he worked on his car every day. Sometimes he worked on it
early, 5 A.M., 6 A.M. He'd rev it up and turn the engine over. Over
and over. Elizabeth became aware of him. He woke her up. She'd
run to the window, stare out, and see him at dawn looking at his
coughing car. Maybe his hands would be tinkering with the car's
insides. Dawn was just another ruined night. Sometimes she'd open
the window and shout: Stop it, stop it. Please. He never heard. He
couldn't hear over his engine. The noise went on and on. Furious
like churlish garbage trucks, incessant like boisterous oil trucks fuel-
ing boilers in basements.

The young super was revving his engine again. No one else
was alive to him. Elizabeth lay there with her eyes open. The noise
grew louder. It always did. She started to inch out of bed. To slide
to the end of the bed. Her toenails were hard. She gouged Henry
on his calf.

What are you doing? Henry asked.

I'm not telling you, Elizabeth said.

Where are you going?

I'm going for a walk.

In the middle of the night?

It's dawn.

Get back in bed.

I can't sleep.

Get back here. Go to sleep.

I can't. He's revving his engine again.

He's got a right to work on his car.

This is a residential area.

What are you going to do?

Tell him to stop.

You're going to get killed.

O.K.

Don't do anything, don't be a jerk.

She might have to die to sleep. She laughed out loud. It sounded hollow in the apartment. She put on her robe and Japanese canvas shoes. Henry pulled the blanket over his head. His back was to her. He'd already accepted her death. Maybe she was as good as dead.

She opened her door and walked down the stairs. The halls were even bleaker in the middle of the night. Dawn. Farmers woke like this every morning, at the break of day, milked cows, sloshed around in the heat or cold, fed pigs who were more intelligent than they were, grew wrinkled and weather-beaten, and their wives cooked heartbreaking breakfasts, shriveled under the sun, nursed belligerent youngsters or died in childbirth. Everyone's a hero. Elizabeth giggled, then stifled herself. There were cigarette butts on the stairs and floors, tissues, candy wrappers, an empty paper bag. Nothing big. No vomit or blood or needles. Only some Philly Blunt tobacco the kids mixed with marijuana.

Elizabeth marched stiffly across the street to the super at his car. She was in her robe, outside, on the street. She knew she looked ridiculous. People do when they act on principle. Like clowns in the circus. She'd only been to one circus. It was a crazy theater, the rings, the animals, the red-lipped clowns hanging from ropes. The audience fears the worst and waits for it. She counted herself a silent, anonymous member of Clowns for Progress. The group plastered its posters around the neighborhood.

Elizabeth stood beside the super until he decided to notice her. She was closer than she'd ever been to him. It was a grotesque intimacy. When he noticed her, she spoke as calmly as she could.

You may not realize it, but some people are still trying to sleep. Maybe even until eight or nine this morning. Do you realize how loud your engine is? And do you know that it's against the law? It's noise pollution. Disturbing the peace. I could call the cops. I won't, but I could. I can't sleep. I can't stand it anymore. Don't you ever think about anyone else?

She stood there. She had finished her speech. She waited beside him, in her robe. He stared at her. His answer was silent revulsion. His disgust should have been reserved for battle, when a soldier calls up the desire to destroy from a vat of disgusting mixed emotions. Pleasure, revulsion, and fear animate the killing machine. Soldiers are allowed legal murder.

The young super, smartly dressed but his nose streaked with

grease, had no understanding of quiet in the morning. No respect for other people who needed their sleep. Elizabeth could see that. She enlivened his killing machine. He and she stood their ground. Her ground felt puny and groundless. They were locked in a barbaric embrace. It was public. They could be watched by anyone. Someone might be videotaping them for a stupid TV show. She was candid and conspicuous. The young super despised her. His rage shaped and reshaped his face. She would've slapped him if she thought he wouldn't murder her. She wanted to wipe the expression off his face. Murder was too good for him. That's what her mother would say. He didn't raise a hand, and the law held Elizabeth's hand. They were both held in check. An abyss yawned, wide and filthy, like a domestic Persian Gulf. She hated her own voice which repeated:

> Don't you understand that there are other people on the block? Don't you understand? People need to sleep. There are other people on the block.

The young super's face had hardened into furious incomprehension. Then he turned away from her, turned his back to her, returned to his car's engine, ignored her existence, and she walked back across the street to her building, walked back up the filthy stairs, went back to her position at the window. Elizabeth wondered who, if anyone, had witnessed the event. A friend or an enemy. Henry slept through it.

That was years ago. Now she wouldn't do that, wouldn't confront someone alone on the street. Now she considered the enduring consequences of announcing grievances to neighbors. Elizabeth had been ignorant of the fact that her super, Juan, had befriended the young super. His name was Achmed, she didn't know which Middle Eastern country he was from, and Juan was Achmed's block mentor. She hadn't known that. After Juan heard about what she did, he was barely civil to her.

What's the difference between Chinese food and Jewish restaurants?

With Chinese food, after an hour, you're hungry again. In a Jewish restaurant, after an hour, you're still eating.

Henry told Elizabeth she had to learn to accept the unacceptable. She tried and slipped and told the woman on the first floor that the woman on the top floor bothered her. The top-floor woman screamed at her boyfriend's child from early morning on, and when she was high on coke, ran out in the night, forgot her keys, and screamed for her brutish mate to throw her a key, to let her in. He'd punish her. He'd pretend not to hear the wailing, piercing shrieks everyone else heard. Finally he'd let her in. She'd whimper all the way up the stairs. Past Elizabeth's door. Then they'd fuck, probably, with the child watching.

Elizabeth complained to the woman on the first floor about how it was driving her crazy. The first-floor woman said that she was friends with the top-floor woman. Did Elizabeth want her to tell the woman upstairs? No, Elizabeth said, no, please. Elizabeth retreated. She had to be more careful. Henry thought she was a jerk. She had to let people know what she felt or thought. He told her she was chronicling her life. He'd watched a TV news special about women talking on the telephone. It said they were chronicling their lives.

The young super never even looked at her on the street. He wouldn't help her if someone was trying to cut her, cap her, molest her. He was an enemy on the block. He wouldn't lift a finger to save her life. In the city, you can have enemies and never see them. It's urbane, humane. But if you have enemies on your block, you can't count on them. Not even in a lethal situation. They might applaud the bad guys or be apathetic bystanders, even grandstanders. Yeah, they could say later, grinning, yeah, I saw him take that bitch and grab her head and slam it against the wall. . . .

Elizabeth fantasized that the young super Achmed would come to her aid. Even though he hated her, he'd help her. He'd overcome his hatred and save her life. They'd forget their enmity, they'd forget the past. They'd become friends, and there would be one less problem in her little world. It was a fairy tale. It was like a dream when an ex-friend appeared and said: I love you. Or something. Elizabeth cried over spilt milk, the irreconcilable.

But Achmed, wherever he came from, hated her. He still hated her. He would always hate her. He still lived on her block. He would always live on her block. Elizabeth watched him. He walked into his apartment building. She hated the way he walked. It was an insolent, arrogant swagger, almost indecent.

She watched the street vigilantly. The street was dead. She was tired. She longed for sleep. The night was long. She was insignificant and small. Sleep wouldn't come. It wasn't her friend. Life was short. Her days were numbered. Her nights didn't count. It didn't matter. No one deserved to sleep.

Barry Yourgrau

Rocking Horse

The bus drops me off in a village in late afternoon. I stand in the earthen lane with my bag at my feet, surveying my quaint surroundings. The little houses behind the piled stone fencings display whitewashed walls and thatched roofs. Their worn thresholds are swept clean. The turf and plantings around them glow startlingly vivid in the moist, windy air. I spot the modest shingle of the boardinghouse, and take up my bag and go trudging over.

The proprietor at his counter is a burly, fleshy type in his shirtsleeves. He cocks his head as I enter, and looks me up and down. I'm wearing large sunglasses, which of course arouses suspicion. As I put the pen down from signing the register, he is smiling at me. "Yes?" I inquire. I blink at him behind my darkened lenses. My eyes burn, as if they were rimmed in coarse metal. The proprietor grins in reply—a cool, open grin, touched with provincial contempt. He has the ruddy cheeks of the locals, the bland, undisturbed flesh of a baby. An insolent, bovine baby, with a lilt in its voice. "Will you be staying just the night, then?" he asks, droll and disingenuous, passing me the key. I grunt an affirmative. As I head toward the stairs, he calls after me, "Our beds're as soft as a mother's lap, just right for—*you know*. . . ." I stop. I turn my head slowly and direct a stare at him. He smirks back, flushed, clearly impressed with the boldness of his coup. He scratches his temple with the butt of the register pen.

I blink and turn away and go stolidly upstairs. I find the room and press the door shut behind me, leaning there for a long moment. Then I hurl the bag onto the bed. I snatch the sunglasses off and fling them after. With an oath I sink down into the rickety chair by the table, and sit with my head in my hands, in abjection and despair.

After a while I straighten, and get blearily to my feet. I rub away at my fiery eyes and shuffle over into the small toilet alcove attached to the room. In the dim mirror over the washbasin, I regard a visage: a haggard face, sparse hair wind-tangled, eyes red and woebegone—and there, on display beneath, the stigmatic shadows. The brand marks of unrest. I probe them with a halting fingertip, as if they were aching bruises from some hideous and abusive ordeal. Which is in truth what they were. I whimper to myself. My fingers trail down my grizzled cheek, and take hold of my slack jaw and exhibit my head slowly left and right. I groan, and put my hands on the washbasin. My head hangs before the mirror.

Finally I splash myself and come back over to the bed. I unlatch the bag and bring out the luxuriant, incriminating muffler and set it carefully on the foot of the bed. I examine again my scribbled instructions, got at such an exorbitant price. I stuff them back deep into a pocket. I tamp the mattress a couple of times with my fingers before sitting on the side. It's soft, as advertised, under the worn gloss of the coverlet. I check my watch to see how many hours I have to wait out. I blow out a long, half-yawned sigh, and blink painfully.

About me the room is almost mean with its drab wallpaper and inevitable insipid color prints celebrating the local environs. A scatter of coal sits in the dirty grate, a paltry icon of coziness. Through the half-curtains, the sky has lost all its color. But the trees close at hand are somehow still luminous, indefatigably glorious. I consult the time again and press on the bedside lamp and sink back in the weak lamplight. My eyes sting. I drift into a groggy, constricted daze.

I twitch, and jolt up. I gape at my watch and run a hand over my face, panting. Someone is knocking on the door. A mocking voice says something about warm milk. "No—thank you," I mumble, reaching over and hurriedly dragging the muffler behind me out of sight. The knocking resumes. "I said no, thank you—*go away!*" I bellow. I glower at the door, hearing the heavy footsteps moving away. I bang my fist on my knee—at the familiar insolence of the taunt, at my handling of it. In my position I simply can't afford entanglements of any kind, altercations. I drag back over to the washbasin tap, my woefulness fully raw and piercing. Self-despising swells over me in a poisonous gust.

Blinking and grim and hot-flushed, I return to the room. Beyond the curtains there is nothing, and still I check my watch yet again, and shudder. I start to pace.

An hour later I descend the stairs. The proprietor pokes his head out of the doorway at the back of his domain. "Off for a nightcap, then?" he inquires, needling. I grunt, an elbow tight against my side, pressing the hidden muffler to me under my jacket as I cross to the entrance. "We'll leave the light on for you—" I hear, almost cackled, as I come out into the night.

It's chilly. Breaths of mist drift by. Up above, the heavens themselves are clear, and the stars swarm in spectacular, transfixing abundance. A night from a nursery tale. I head left along the lane, per my instructions, passing cottage windows aglow in the darkness. The light from them is candle-soft, an aching tenderness prefacing slumber. I grit my teeth and quicken my pace, trudging along. The sounds of carousal come to hearing—the rough harmonizing of a snuggery. A wedge of light spills from a partly ajar door ahead. The clamor of voices grows louder, maudlin and disorganized, but still famously lilting. I draw closer warily. A burly figure lies slumped against the foot of the wall by the doorway, his bare head thrown back, drink-blinded and snoring, in a gross, bestial version of sleep. As I drift past, I stare at him—at him, and then hurriedly beyond him, into the glowing, forgelike depths of the snuggery, and its roaring mawkish choruses. Two smoky figures come tottering out suddenly into the lane, staggering this way and that, arm in arm in gargled song. I rush away from them wildly. When I finally look back, I see them swooned in a heap on the ground.

I reset my muffler under my jacket and put my head down and renew my pace, my heart thudding. I tramp along for several minutes. The lamplit windows become intermittent, and then more intermittent. I stop. Panic laps me. I probe the night all about with my burning eyes. Up ahead, there seems to be a faint light. I edge toward it. A thin black shape takes form at what must be the mouth of a smaller lane, per instructions. A figure with a flashlight. I call out the password in a constrained whisper. I receive the reply. I hurry forward, as my guide-to-be flaps at me to approach. "You weren't followed?" he demands. I shake my head. He's a gaunt, unappetizing creature, the telltales showing under his own eyes, I notice with disgust. He leans at the waist and peers back down the way I've come. He looks up at me. "All right, then," he says, and he grins with presumptuous collegiality.

I follow him and his low bobbing light down the rutted track. There are dark blanks of fields on either side. "So what do you think of our lovely country, then?" my guide whispers back, in an inane gesture of hospitality. "Is it much farther?" I reply, irritably, ignoring the query. His ankles are bare and scrawny, pathetically sockless under the too short lengths of his pants. As he starts to answer, I freeze. "What's that?" I cry, my heart erupting in me. "What— what?" he cries. "Over there!" The flashlight beam swings up into the field. "Aw, sweet mother's milk!" my guide exclaims, his voice scornfully easing. The cow blinks at us in the light. My companion cackles, shivering with chronic fatigue. "You put the fear of God in me like that," he rebukes. "I almost dropped my torch!" "All right, let's go, let's go," I say. We resume. "It's just along here, I'm telling you," he declares. I don't respond.

Finally a patch of muted light gleams. A curtained window shows ahead. "You see? Home, sweet home," my guide announces in a whisper. My heart rises up again. We approach slowly. I'm made to wait a few yards off as my guide goes up to the low door in the freshly whitewashed walls. A chimney juts against the heavens from the thatched roof. The gaunt figure taps a sequence on the door. "It's us, grandmother," he calls softly. He presses his ear to the planked wood. He looks up toward me, grinning, and waves me forward. "Go on in, the dear old one is waiting," he declares, in a cloying singsong. There's a pause. He remains between me and the door, grinning. I pull out some loose currency and thrust over a crumpled note. He winks and steps away. I feel the muffler with my elbow. I tap uncertainly on the door, and then again, until a low, weak voice answers, and I go in.

A small elderly woman stands in the middle of a one-room cottage. The cozy warmth engulfs me. She nods several times, awkwardly acknowledging my equally awkward greeting. She is the authentic, sweetly wrinkled article as promised, her countenance still untrammeled, her gray wisps of hair trimmed under a kerchief, a black shawl draped over her stooped shoulders. Her tender eyes flash with unvarnished cupidity as I pass over a thick wad of notes. She indicates the bed in the corner with an ancient, grossly be-ringed hand.

I go over as bidden and sit stiffly on the edge of the bed while she busies herself in the depths of a cupboard. The bedstead is spanking new, as is the garishly tasseled coverlet. But the rocking chair by the pillows is discolored and loose-jointed from use. A tool of the trade, the illicit fruits of which are all around: the electric faux-turf fire in the hearth, the enamel fanfare of the kitchen area . . . the festooned, splendiferous grandfather clock, the baroque-fringed bedside lamp. A bumpkin's idea of corrupted wealth. I blink in the artificial heat and stifle a taut yawn.

The old woman finishes her business and approaches. My heart resounds. She smiles down at me, nervous, showing false teeth. "Shall we begin, then?" she says. I nod, staring off at the linoleum floor. I'm trembling. A wave of sheer craving swamps me. I pull free the muffler and, gasping, I wrap its soft mass about my neck. "Here?" I murmur in a quaking voice, looking about at the pillows. "Yes, yes, love, go on," she mutters. I have a brief, ugly vision of myself—a shadow-eyed monstrosity being serviced by an avaricious crone. I sink down into the aroma of linen. The low, ersatz firmament of the ceiling hangs over me. The rocking chair creaks beside me. "Hurry—" I murmur frantically, closing my burning eyes. The creaking repeats, and slowly starts a rhythm, as the old woman begins to rock. Syrupy warmth seeps around the knotted core of myself. A tear wanders out from the clenched tip of an eyelid. I moan, and roll away on the pillow, and she starts her lullaby.

Her voice is self-conscious and off pitch and thin. But it has the required lilt to it, as it warbles the prized intimacies of the cradle song through the creaking of the rocking chair, through the clacking of false teeth. I groan low and shudder in my muffler, pressing my hands in between drawn-up knees. A nectar loosens wide open in me, and oozes into every last corner of tissue, like a thick-flooding illicit honey. It's sleep, sweetest, childlike sleep, the purest and finest of slumbers.

My eyes jumble. I wriggle in protest. I blurt a groan, aggrieved, and my eyes wrench open. There's a clamor in the room behind me. Gasping, I swing in groggy chaos toward it. The rocking chair bobs back and forth, empty. The old woman is in the middle of the room, a hulking constable towers over her, the door is ajar to the darkness beyond them. The old woman simpers about "just a sweet young friend, just come along for a visit." She has a hand to her mouth, to screen her dentures. "Now, I wasn't born yesterday, you know, grandmother," the constable keeps rebuking her, shifting from one boot to the other.

Finally he comes trudging over toward me, like a brutish giant from a children's fable, splaying shadows over walls and ceiling. I feel the familiar prod of a truncheon, hear the ugly names, the sprayed order to clear out. "But I've done nothing wrong!" I bleat. "What's that for under there then, eh—eh?" he demands. I wince at another hard gouge. "Corrupting elderly gentlefolk for your sordid ends!" he bellows. He lifts his truncheon, as if to strike me. I cringe. "Clear off, you—!" he cries. "Clear off!" He grabs a handful of my jacket collar and hauls me to my feet and goads me toward the door. The old woman stands off in silence by her ill-gotten stove, her gaze fixed on the glistening floor there. The constable rebukes her one final time, shaking his head, his tone almost heartsore.

We come out into the chill darkness. My onetime guide is of course nowhere to be seen. I stumble along, head bent beside the constable, who wheels his bicycle. The bell jingles as the track bumps up and down. The headlamp flickers. The constable's eyes are small and inflamed, his features blurry from sleepiness, his hair disordered, as if he'd been roused from his snores. "You foreign lot," he grunts in contempt, his breath fogging. "Why can't you make do with a pan of warm milk? Why can't you take liquor, then—like a man!" I don't answer the old, scornful charges. I blink, and swipe at my painful eyes. Sodden wakefulness sits like ashes in me.

At the crossroads the constable sends me on my way with the common snarled warning, to be on the first bus—or else. He lurches off for bed on his bicycle, like a costumed barnyard animal attempting a circus trick. I turn away, and plod back down in the direction of the boardinghouse. My spirit is broken. Once more I'm out of pocket, reviled, awake as ever, cast out from access to that one sweet simple thing I crave above all else. A sob drifts from the deepest part of me. Above my bowed head, the splendid heavens swell, like the mockingly displayed wealth of all the dreamy El Dorados of sleep.

Paul West

B UYING THE FARM

At Starfish Speed

Booth and Clegg, who had been military pilots together, transferred to the airlines like two golden workhorses going to heaven. The martial rigor of their manner remained, but each had learned to nurse the other's needs. A precious decorum had grown between them, from such a phrase as "starfish speed" (meaning slow) to a forty-five degree bow that indicated apology. The phrase was a weakling of a code, the bow was vaguely Japanese. When they needed to speak of themselves combined they referred to themselves as "Miles Magister," who, mainly in their Air Force days, became a real person to them, and just vaguely so when, after a short flight of an hour or so, they parked the DC-9 and limousined off to Holiday Inn or Ramada, sometimes even a HoJo or a Travelodge.

There, side by side on their double beds, they became a supine variant of their cockpit duo, but sometimes thought Miles Magister had signed for the room instead, relieving them of the burden of identity. One of their milk runs was from Pittsburgh to Ithaca, departing at 9:15 in the evening, arriving around 10:00, but almost always late because this, the last flight to Ithaca, waited for "runners" from California flights that got in around 8:30. After so much adrenaline spent, they found it hard to sleep, Booth more than Clegg, who managed to drift off only if Clegg read to him from a volume of accident reports, or read to him such items from a magazine, somehow banishing the ghost of what might have been by insisting on the worst. Or, sometimes, Booth read to Clegg until the book or magazine fell from his hands. Their mode of reading was casual, requiring many paraphrases of the boring parts, omissions and occasional elongations of matters too juicy to be read in their own sweet size. Now and then either of them found dramatic oral exchanges between tower and pilot that required histrionic emphasis and, after that, a short pause.

Tonight, Booth and Clegg, Barney and Rupe, were in a Holiday Inn, locus of the fabled King Leisure suite.

There was one history, of a young Mooney pilot who left San Antonio in weather conditions clearly beyond his skills. Yes, he had told the Flight Service Station he could go to instruments if he had to (and he would *have* to, they told him). He spoke as if he did not understand instrument flying, insisting on a route he called "080 to destination Clover Field," which was not the way to say or do it. It was suggested he accept another routing, of Vector-198 to the Eagle Lake VOR and then direct. A note of derision and chagrin crept into Clegg's voice as he went on with the reading while Booth lay back and tried to envision tragedy in the making.

"He loaded his wife, mother-in-law and infant child into the Mooney and requested clearance at 4:38 in the afternoon. *Uh,* he said, *we've filed our plan and, uh, uh, uh, need a course of 082 to Eagle Lake and then 091 to, uh, Clover Field south of Hobby.* While he kept asking for what he wanted, the clearance-delivery controller kept trying to give him the course he needed: *V-198 to Gland intersection, direct Hobby, direct,* as well as the usual altitude, frequency and transponder code information. Oddly enough, the frustrated controller did not insist on a read-back but let him tootle off with the word *copy.*"

Booth, who had heard this one before, but responded to its retelling like an aficionado to part of a certain operatic aria, doting on the outrageous finesse of it, broke in and said from within his narrowing tunnel of sleep, "Don't tell me. He took off in appalling conditions from San Antonio, with a ceiling and visibility so poor the tower controllers couldn't see runway or taxiway. Then he said *rolling,* and took off into the murk. Rupe, tell me the bit about the controllers, when they talked about him like the bad boy in school."

"The tower controller asked his colleague in the radar room *Departure, did you turn 81K?* And he answered *I'm not talking to him.* So the tower controller called the Mooney pilot and requested his heading, only to hear *81K, uh, needs a course of 082 to Houston.* This pissed the controller off, so he issued an order instead, saying *Mooney 81K, you fly a heading of 120,* using the good old imperative and don't you jackhoss me anymore. There was no answer, and Mister Uh-Uh-Uh, he flew off, closely followed by a Piper Cherokee."

Booth half-woke and said with jubilation, "Now it gets interesting. They know he's on 120 but they don't know if he's there by accident and they turn the Cherokee to 180 to avert a collision from behind. In the confusion—" He yawned hugely.

"In the confusion," Clegg resumed, "the tower controller mixes up the numbers of the two planes and the Mooney guy answers something intended for the Cherokee. Imagine! The Mooney pilot never responds again, no doubt in an all-befouled-up sulk. Back in the tower, conversation rages, with the departure controller saying *What's 81K going to do?* Then he asks the tower this pathetic old question: *Are you sure he's even airborne, Bobby? Yair,* comes the answer. *There's not a target or anything out there now—there's not a airplane, there's nothing out there.* Truth told, that Mooney was all over the place until the right wing tip hit the ground. The fuselage went between two big mesquite trees, losing both wings. And the cabin with its passengers flew another hundred yards until it hit a substantial tree trunk. No fire. All dead."

"Recrucify the Infant Jesus," Booth droned, "you wouldn't get me taking off into conditions like that: 200 ceiling, visibility half mile. No way. What puzzles me is, he *knew* he knew nothing about instrument flying, but he was determined to go. Maybe to show off. But why should he think he could get away with it, less than a novice, when experienced pilots wouldn't even attempt it?"

"He was all balls, Barney. They often are. They spend a quarter million bucks on a deadly toy and then they prove just how deadly. I guess somewhere in the mental procession from purchase to death there happens a moment of sublime heroism, at least as the

guy sees it, in which he is giving himself a chance of doing something almost impossible, and then he finds he has the balls to keep
his lunch down and his bladder tight and his anus puckered while
doing it, after which he knows he is still alive but for not much
longer, maybe a minute or so, long enough to mull the matter over
before he hits. I don't know, Barney, there's more balls and bucks
than there is brains." Clegg saw himself back in Turkey, holding
forth in the officers' club, opining about flight at 80,000 feet, about
which he knew too much. They listened to him there, at least until
Booth entered the room.

"Let's assume," Booth said dreamily, "this guy wasn't trying to
wipe out the family in one fell swoop. So he must have thought he'd
figured it out for himself. Look how he insisted on his own route,
like somebody downtown asking a policeman. *Telling* him. The auto-
pilot would have done it for him, but what did this fuck know about
autopilots anyway? His VCR must have plagued the jissom out of
him, not to mention the can opener. He became disoriented almost
at once. He was an amateur trying to do a professional's job. I'm
glad he isn't in the sky anymore, ready to kill us all." That was
enough. The drug had worked, the enemy had been routed. Death,
on whose side Booth so often found himself, had won again, and for
good reasons. It pleased Booth to think there was logical justice in
the sky, keeping the good guys alive and the dunderheads in the
ground. Clegg felt some sadness, a streak only, but wished he did
not have to recite these accidents to the somnolent Booth, to whom
violent death was a specialized pornography; it made him come in
his mind and gave him a fair night's sleep, and perhaps he would
have enjoyed photographs too, a smell at the ripped clothing, the
sundered fuselage. Pilots could become so good they felt callous
and, when some poor inexpert slob bought the farm, forgot the

reverence for magic in his woeful performance, the tender openness with which he started out to get his ticket. Booth joined forces with the headsman. Had he not, after all, survived his own military mis-adventures, for some of which he'd been responsible? He must have felt invulnerable, even in sleep, Clegg thought.

Why does he like me to read that other one in a high voice, the bit by the woman in the crashing Bonanza when she says *going into the mountains*. Her last words. What's the charge for him in that? Why did she bother saying it? Perhaps to adjust her own mind to the unavoidable possibility that it was all over for her and the mountains were her last, last thing.

The Falcon

Yair, death, Clegg told himself, death in comparison with which everything is interesting. Imagine being close to that stultifying comparison. It was as if his thought awakened Booth, who had seemed in a deep slumber, half snoring. There were his eyes, negat-ing the sleep already had, and he was asking for the *Falcon* in Lake Michigan. Clegg was glad; he could do this one in short order and nod Booth off again with a précis. There was no need to read it out loud; after all, eleven o'clock was coming up and with it *Un-solved Mysteries* on the TV bolted high above them like a railroad signal.

"Okay, the *Falcon*," he began, Booth eyeing him in a manner close to hostile. Get on with it, Booth's glance meant. "The cause of this one was the parking brake in the Falcon 10. With three detents, it had to be dealt with carefully. Off. Park. And Emergency. If it's not Off, a light low down on the between-seats quadrant shines to warn you. The pilot claimed he had released the brake on departing the parking area and had not used it subsequently while he was taxiing or while waiting to take off. However, investigators found, and this is what sickens you, the *Falcon 10* taxis quite well with the brake set in Park, and detailed studies of the brake handle found in wreckage proved it had been in Park on impact. Then they looked at the brake pad material, which had been overheated to something extreme, and then suddenly chilled, as would be consonant with the *Falcon*'s plunge into Lake Michigan. So it took off, but not airworthily, *with its brakes on.*"

"Can you believe it?" Booth asked. "Talk about a five-million-dollar toy. I have heard of guys taking off with the tow bar attached, or with the elevators locked, but this—now tell me the part about the crew."

"The *Falcon*," Clegg said without referring to the ripped-out article, "floated for a while before going down in twenty-five feet of water. Not too bad, I reckon. Depends on who you are, what you're made of. Scratch the copilot, who died of injuries. The pilot swam out through a gash in the fuselage bottom made by the entire nosewheel assembly being torn out. Lucky guy. As for the four passengers, they went through the overwing escape hatch and treaded water, held on to floating seat cushions and actually got on floating hunks of ice. A fire department helicopter was able to fish them out before they got too cold. It's obvious, Barney. The *Falcon* should have aborted, even if it still hit the lake; it wouldn't have hit it so fast or gone out so far. Instead, they elected to take off, impeded, and went nose-first in."

Booth had drifted off again, partner to those five in the water, dreaming his down-in-the-drink dream, mourning the costly jet. When they died, he knew, they all died together, the expensive like the cheap. It grieved him to have a complex airplane bite the dust or the water; he thought of all the welding and circuitry, the formers and electronics, devoted to making it behave as required. Then some yahoo came along and made a pig's ear out of technological poetry. In his dreams he sided with the Germans, who, although often wrong about air strategy (too much emphasis on versatility; the cult of the two-engined bomber), were technical aces, always making the engine accessible, and the guns, and forever packing the wings with automatic slats that kept the plane from stalling. Air pressure kept them within the wing's contour, but out they popped as soon as the airspeed fell toward stalling. In another life he would have been with them, shivering on the Russian front outside his *Arado* reconnaissance plane, knowing all was lost. Perhaps he felt for the Russians because the Germans were so good.

Against his better nature, Clegg had fallen to reading yet another of their bedtime stories, the one with the woman saying (in an earlier transmission), "Bonanza 725, we're in trouble, we're losing airspeed fast, going into the mountains." That was at eleven twenty-five and thirty-three seconds. Only three and a half minutes earlier all had been going quite well, and the exchanges with Oakland Center had been mellow, composed, patient:

"Bonanza 725, ah, how's the ice on the wings now, sir?"

"725, ah, ah, a little more icing."

"Okay, is it still building up or does it appear that it slacked off now?"

"Seems to be staying on the wings. 725."

"Okay, sir, are you getting any more?"

Clegg shuddered, noting the minimal passage of time between "on the wings. 725" and "getting any more?" Only seconds, and the amount could increase even as you watched it dust you down.

"725, negative." Clegg sighed relief and went on reading,

holding the magazine far from his eyes as if to avoid contamination.

"Soon after this the woman took over the mike and loggers working northwest of Redding, California, saw the Bonanza appear from the clouds and, with a revving up-and-down noise, seem to spin, blast into the ground at 3,300 feet, to be followed only seconds later by a ruddevator that pinwheeled down from the bottom of the overcast. That was the left ruddevator falling; the right-hand one had been dragged down after the rest of the airplane, still being connected by control cables." Clegg tried to imagine the pilot's wife and her state of mind, knowing she had taken over the mike in the final seconds of their lives, as if she, the nourisher of life, might call on some prevailing force to save them at last. In fact, Center's last question, to the pilot, had gone unanswered. "How are you doing now, sir?" brought no one into action aboard the Bonanza, so Center made a longish suggestion: "Suggest you, ah, descend now and maintain niner thousand and, if you can, reverse course and proceed back southbound." Then just the woman and nothing, the pilot having lost control of the airplane.

Clegg was mighty glad that the controllers called you "sir," suggesting a courtliness or knightliness of the airwaves, an exaggerated courtesy that as often as not fortified a nervous pilot and, merely because he was being addressed in a respectful manner, made him find courage he never knew he had. He became a hero because he was being addressed as a gentleman by a gentleman who probably had no such elegant motives but said "sir" out of mechanical good nature. Clegg liked the "uh"s and "ah"s as well, surmising they were indices to honesty. People were thinking under pressure, not always able to produce the perfect answer at heartbeat speed, and indeed sometimes pausing before the most ordinary word or

number. Chivalry, he decided, calms the heart and stills the churning bowels. We have all been there, our mouths too dry to talk with,
our hearts doing a jackhammer in our chests, and then some total
stranger out of the uncaring invisible ether calls you "sir," and you
feel knighted. Surely that was the way, even with student pilots:
strength through civility was the proper motto. No wonder he still
called Booth "sir," not that Booth needed buttressing; he had too
much courage as it was. No, Clegg enjoyed the sense of collegiality
the finesse brought, the sense that they were among the elect and
used little pomps that other men could not rise to without seeming
effete. Envy not the controller, duty-bound to turn novices into
master pilots with a little touch of patrician calm. In such a world,
Clegg thought, it was a wonder anyone crashed. He had missed
nearly fifteen minutes of his TV show.

Gulliver

Yet tonight was one of those nights on which Clegg was unable to
concentrate on the haggard, whispering narrator who always
seemed to know more than he confided, or on the simulated faces of
the lorn, the bereft, the permanently upset, pictured on the show in
party frocks, baseball suits, PFC uniforms, wedding dresses and tux-
edos, all of them with lives gone palpably wrong for long or short
spells. This was the texture of life lived next door to death, sometimes put right by a phone call; but even the life's presence on the
show did not guarantee it irresistible allure. It could be dull, tedious,
durably generic. All the same, even the dullest life seemed an object
of ceaseless fascination contrasted with death, about which, Clegg
reminded himself, we knew nothing at all, even less than the Higgs
boson, about which there had been so much debate in scientific

circles—what was it, and why? Clegg had no idea, but he knew when things had no qualities, and death was one of those things. He understood why Booth doted on these gruesome bedtime stories, reassuring him and making him feel he had done a great many things right amid the mayhem of his calling. *There*, Booth would keep thinking (or so Clegg thought in his Euclidean way), *but for the grace of God go I*. He died in every fatal crash, emerged smoldering and bloody from every nonfatal, and actually seized the controls in some, saving all souls aboard like Gulliver leaning out of a windmill and lifting Alice from the bottom of the well.

"Look, I *gotta* read you this," Clegg said. "There was oil pouring from the oil access door on the right nacelle, see. The dipstick was gone. The number five piston had a hole burned in it. And there was burn damage on two connecting rods, and the rod bolts on six had failed. This guy had eight hundred hours' total time. He was good enough to close down the malfunctioning engine and make the ILS approach single-engine. He just panicked out when he could have made the decision to shut that engine down and work with that situation."

"You big-league junior piss-cutter," Booth snapped at him, "leave him alone. The guy is dead. He ruined his family. I want to hear what one of them said. Wasn't it the woman who survived?"

Clegg read, slowly and earnestly after the rebuke. *"I was just so scared I put my head down and I was holding on for dear life and they were screaming and saying we got big trouble, big problems, big problems, and Larry was screaming something and Steve said oh my God I can't believe it, oh my God. . . . All of a sudden we hit some really heavy turbulence or something and our altitude dropped and we were going down and all of a sudden the plane just went crazy. We just lost control and started diving. First we went straight up and then we started going down. I was looking out the window and I saw that we were going down straight at the trees and then we leveled off right before the trees. And that's it, that's all I remember."* What a silver survivor.

"Yes," Booth said gently. "That's our stock-in-trade, Rupe. You and I are two guys who dominate that emotion and live to tell."

"How come, then," Clegg answered, aware of deviating from some norm even as he said it, "it don't feel like survival? It feels like we went down with some goddamn ship. Honest, sir."

"Surface feelings, Rupe. Deep down we are survivors."

Clegg could not quite muster the sense of triumph; he felt mesmerized and lethargic, no doubt a result of reading accident reports aloud to a man who appreciated them too much. Would they have fared better if he'd read to Booth from *Hiawatha, Dracula* or *Treasure Island?* He groped for favorites, but concluded he hadn't read many books, or couldn't remember those he had. He had read his share of Eric Ambler, Leslie Charteris and John Buchan, but the only titles he could recall were not by them at all but by more demanding authors, or so he guessed: *Gulliver's Travels, Crime and Punishment, Erewhon.* And there was one flying novel that had held him, *Pylon,* and one about a German U-boat called *Das Boot.* He wondered why there had to be German in the American title, but supposed it was a matter of flavor, something lost if you said *The Boat,* which could have been about Noah's Ark. Clegg's relationship to literature resembled a free-falling parachutist's relationship to given points on the earth's surface: the Empire State Building, Dutchess County Airport, Lake Cayuga right beneath him: uncertain and tumbling, prey to extraordinary shifts of attitude and trajectory. Usually he liked reading to Booth, but there always came a point at which he, Clegg, wanted to throw up. No one would realize it, he thought, but there were planes tumbling out of the sky all the time. Never a shortage of disasters to fit the accident columns in the magazines. Not commercial carriers, not military, and not business, but private or what was known as general aviation: amateurs, really, who could never get the flying bug out of their systems. They flew into lousy weather, stalled and spun, ran out of fuel, did crazy stunts to impress their relatives or women friends, and that was that: an expensive toy totaled in minutes. He sympathized with aerial yearning, but deplored the feckless rodomontade of the pilots.

The Navajo

Look at the guys in the *Navajo*, the subject of his most recent elocution in that graveyard voice of his. A shot magneto had kept them on the ground for hours. They were three guys heading for the Bahamas. They left Long Island after midnight, reached Florida in time to meet a tropical storm. When they took on oil, the left engine took seven or eight quarts, the right one three. Talk about indices to bad fortune. When at last they began to taxi, the *Navajo* went off the taxiway onto the grass. Then followed the scenario he had read to Booth, graced (as he, Clegg, saw it) with such mellow formalities as saying "with you" to the tower, Departure saying "all right, sir, say intentions," as if nothing bad was going to come of all this bungling and bad luck. They all called the pilot a squirrel anyway, so they had an attitude from the outset, which must surely have worsened their attitude to him as the day wore on. They called him "sir" all right, but underneath the amenities there was this tone of pawnbroked respect. Tower complained to Departure *It took us 30 minutes to get him on the runway. Uh, uh,* as Tower would say. Then, at a later point, Departure seems to whisper gossipy fashion *Hey, how do you think I felt?* And Tower answers *Okay.* Around this faltering guy has grown a susurrus of incipient slander that maybe did not affect him, though it easily could through timbre and intonation, whispering behind his twitchy back. Then Approach tries to say something, but it comes out *He's gonna be, gonna be, he's gonna be . . . after Osprey 152.* They were wondering where he would fit into the conga line of planes coming in to land at Jacksonville (JAX), their only duty really being spacing and separation, although they did many other voluntary things as well. *I'm gonna call Alert Two,* Tower says. Approach, almost chidingly, tells the pilot *Fly the best 290 you can, sir.* The pilot answers, a bit crestfallen, *Roger, I appreciate any help I can get,* little knowing he has only a couple of minutes to live: *coupla,* as they say in the trade, and they did not go to the Bahamas at all.

"Roger, sir," Clegg whispers to someone, appalled by something so uninterrogably simple as a fatal accident, selective in that two survived. He would much rather they chartered a plane and allowed him and Booth to fly them, at starfish speed, to where they wanted to go, with affable banter all the way and not a whitened knuckle among them.

The amazing thing about Booth and Clegg, who by now had racked up as many hotel hours as flying ones, was that they found nothing incongruous in repeatedly using death and terror as a sleeping draft, though one not that effective. Perhaps they were doing a voodoo on bad luck, making a boon or a grace of the thing they dreaded. Booth rarely slept well, though Clegg slept better if he'd been able to watch *Unsolved Mysteries* in peace. Usually not: Booth, once a light colonel, required Major Clegg to read to him, like a standby teller from the *Arabian Nights.* He was the caliph and Clegg was the houseboy. One day the airlines would separate them forever, turning Booth into a flight inspector with power to crush the career of any Clegg, promoting Clegg into a Booth. Not all their nights are the same, of course, but the readings come from the same yellow-leather attaché case that Clegg carries about with him. A night no sooner begins than it merges into all the others in which Booth mortifies himself, Clegg assisting.

Symbiosis has no finer pang unless, in a similar room, King Leisure or not, a couple of other jet jockeys soothe themselves to sleep by reading aloud an account of how Booth and Clegg screwed the pooch and bought the farm by staying up too late.

Luisita Lopez Torregrosa

Insomnia,

a Romance

A friend of mine who's had insomnia since she grew to know fear and dreams and loneliness, since she was, say, eight, keeps a notebook at her bedside to fill the hours, and when that fails her, she wanders around her house, room to room, plucking dust balls off the carpets and pumping the cushions. On those nights when she hits the wall, when neither ruminations nor perambulations bring sleep, she'll drag out the vacuum cleaner and run it over the lampshades, but most often she sits in her living room in the dark with her cat, the running motor that hums in him finally bringing relief.

It was on a night such as that, years ago, when her marriage was dying within her but she didn't quite realize it, that she left her bed, in bare feet, in her pajamas, in the winter. She sat on the porch of the house she had then, with a dog she adored, and watched the dark turn to silver, and knew then that her dog would die and that she would leave her husband.

Revelations are, of course, the stuff of night, and sorrow, despair, dreams, passions. We all have bad nights, our dark hours of the soul, heart palpitations and chilled sweats. But insomnia is special, perversely romantic, addictive. It lets out the demons. It brings brilliance and melancholy, childish terrors and wicked fantasies, riffs of music and words playing incessantly in our heads, nights racked with anxiety and grief and longing.

On the night of the day my mother died, I lay in my bed, dead myself, and the walls seemed higher and whiter and empty of shadows, and I left the bed and went to my living room and sat up all night, eyes grained, tearless, as if I were waiting for somebody to come and tell me that it wasn't true, at all.

I went to my sisters the next day, flew halfway across the country to the place where they live, where my mother died, leaving us too abruptly, without advance notice. There had been no time for bracing and smoothing my face and binding the wound. I was bleeding profusely, but I saw no blood. Sleeping pills, Valiums, herb cures were offered by my sisters, but I refused. I would sleep unaided.

I stayed in a motel near my mother's house, near the funeral home where her body was on display, rouged and coiffed. I had refused to see it. It was not she. That night I went to bed, tearless still, alone in a room of double beds and cable TV at a roadside inn. I knew I'd sleep, after all that had happened, but sleep didn't come, not for an hour. I read the local papers. I turned off the lights. I saw my mother, in her short red velvety robe, drinking her soporific, a glass of milk before bedtime, and I saw that house of hers, where the fruit in the bowl on the kitchen table was a metaphor for her last years—wax apples and pears for a woman who grew up with ripe plums and dripping guavas. Her fine-china espresso cups, which she had packed and carried from one country to another, and then from town to town, from sand plains to cactus hills, gathered dust in the cupboards, where she kept them for the long-anticipated right occasion, an evening of great chatter and piano playing, but that, too, had not happened to her.

In the morning, when the light finally came through the thin motel curtains, I roused myself, showered, dressed in black, and went to the motel front desk for a cup of stale drip coffee. The others in my family were beginning to stir in their rooms, well slept, rested, ready for the rituals of burial. Who would sit on the front chairs under the green tent? Would we go to the lunch the church was giving? Would we pose for the big family photograph—a family that included a few blood relations and a large contingent of step-everythings, and town folks, people who called her by a name that I didn't recognize as hers, who saw her, not as I did, but as someone I didn't quite know.

For weeks after that hot, cloudless September day, I did not sleep. People who had lost a parent warned me that she would surprise me, appearing suddenly, as if for a visit. That didn't happen in the daytime, when I was taken up with my small routines, but I would stay awake at night waiting for her. Her face had no lines, her eyes blazed at me, as they had when I was a child. The hem of her dress came to her knee, a slim skirt belted tightly in wide patent leather, and her blouse, with its intimation of cleavage, contoured her breasts. She liked her figure, that she had a flat stomach (actually, there was a soft roundness to it) even after having all those children. On those visits she sat in my armchair, her legs crossed at the knee, showing off her perfectly shaped legs, sheathed in nylons, and her small feet, always in heels. Her hair, fine and dark, grayless until she was quite old, fell in permed waves around a face that was oval and Romanesque, classic Latin, at times forbiddingly beautiful.

She asked me questions that I would answer with no thought of sparing her, and I listened to the illusion she made of my father, who broke her heart, and the illusion she made of her present life. I wanted brutal honesty from her, but she needed her dreams to survive, and I was too selfish and angry to give them to her. Sometimes she cried, sometimes she sat quietly, lost in thought, polishing her nails lacquer-red, the cuticles bitten raw.

Less often she made her appearances in my nights of insomnia as the older woman she was before she died, totally gray (she refused to color her hair), her skin fragile, reddened. She still had her figure, and in the last picture of her, taken only a month before her death, she looked splendid, with a new short haircut and an above-the-knee black dress, at a candlelit table in New Orleans.

I think I know now why I kept myself awake. I wanted to see her clearly, and only in the palpable solitude of night, when no other visions or voices could distract me, could I give her the time that she had so much wanted with me. Only then could I truly hold her to me and keep her alive.

. . .

We all know the sleeplessness of love, the sex of fantasy, and the sex that leaves us longing, unsatiated. The ecstasy of it, when we surrender all, and skin, bones, and blood flow one to the other, a rushing fluid. Our days and nights of somnambulism, when we are left unmoored, transported somewhere melting and flaming, and sleep cannot possibly come, or refuge.

In the autumn of a year when I fell in love, I would stay awake, lights out, eyes shut, waiting for a phone call. When the call came, a voice, over hundreds and thousands of miles away, ran over me like water, like air, making love with words, with silences. We could stretch the night into morning, and then, exhausted, we would part, but I would not sleep. Insomnia was romance.

Our nocturnal raptures gave life to fantasies, the most daring images, things we would not utter in daytime. Darkness sheltered us, and distance, the safety of it, freed us to abandon the usual limits. Afterward, I would get up from my bed, or the sofa, and poetry flowed, or so I thought in my stricken state. I longed for that touch now cut off, unreachable. I wanted to make the submersion last. Staying awake prolonged it, pushing away the inevitable moment when I would come to my senses.

There are those who fall asleep immediately after sex, for whom sex becomes the ultimate cuddling, a sweet passage to subconscious dreams. But those like me, passionate insomniacs, for whom tranquillity is alien, prefer no endings. In the full grip of the madness that is falling in love, I never want sleep, blankets over us, breakfast in the morning. I want never to stop, and on those occasions when it has to end at some point, I leave the room and do the worst possible thing: I play Beethoven. Not his quartets, not his bagatelles, but the symphonies, big, swooning, the heavens opening.

For me romance is deadly. I can never balance the routine and comfort of love, fond love, as Iris Murdoch calls it, when one sleeps soundly, against passionate love, which, as she has said, always means trouble, tumult, insomnia.

"The nights when things end. Those are the interminable nights," Graham Greene wrote, summing up.

The day our love ended, broken, unrecoverable, when its tired beauty and lingering tenderness could no longer hold us together, I went to my bed, turned off the light, and pulled up my old quilt, and I shook all night through.

For a very long time afterward, living alone, unconnected, I would stay awake night after night, living it again, sinking into that black hole. I played music and stared out my window and drank wine and cried torrents and clutched the cushions. Sleep came only in spurts of tossing and turning and jolting up in bed at four in the morning, thinking my lover was in the room, or worse, knowing she wasn't.

But now I have mastered insomnia. If I can get myself to bed before midnight, if I don't drink, if I lull myself with a safe book, say *Sense and Sensibility*, but definitely not the poems of W. S. Merwin, and if I can keep from burning up in flames, which is what happens when I listen to Albinoni, and if I drink water and take a couple of

Tylenols and the phone remains silent—if I don't have a deadline to meet or a romance in mind, I have a shot at sleep.

Then all I have to worry about is sleep itself, and the dreams that come with it. And I have to wonder if it isn't better to stay awake after all.

Quentin Crisp

Insomnia

I live on the Lower East Side of Manhattan, on Second Avenue just beyond redemption. I share a sordid block with the Angels, who drive their motorbikes up and down the street throughout the long, dark night to the annoyance of the neighbors, who complain of insomnia caused by this activity. I have pointed out to them that to object to noise in a big city—especially New York—is ridiculous, that what keeps them awake is their indignation that anyone can be so thoughtless.

When you go to bed, resolve all anger, fear, or frustration. Indeed relinquish all thought. The absurd device of counting sheep is merely a transparent ruse for preventing you from thinking about anything important. Lie in bed, flat, spread out, and breathe deeply through your nose. Concentrate on your breathing, and if any thoughts other than of this activity enter your head, dismiss them. Live inside your body in the continuous present. Do not worry about what you did wrong yesterday or will almost certainly do wrong tomorrow. Lie still. Do not look at the clock. Shut your eyes and next day you will feel as rested as if you had slept.

Napoleon only slept four hours a night. You should worry.

Of course, it is easy for me to give this advice. I sleep all the time. I am what is grandly called narcoleptic. I sleep in the cinema, at the theater, in my room while watching television, or indeed anytime that I am not walking about. Perhaps that is why I have lived so long—and why I don't rule the world.

Anton Chekhov

 LEEPY

Night. Varka, the little nurse, a girl of thirteen, is rocking the cradle in which the baby is lying, and humming hardly audibly:

> *"Hush-a-bye, my baby wee,*
> *While I sing a song to thee."*

A little green lamp is burning before the icon; there is a string stretched from one end of the room to the other, on which baby clothes and a pair of big black trousers are hanging. There is a big patch of green on the ceiling from the icon lamp, and the baby clothes and the trousers throw long shadows on the stove, on the cradle, and on Varka. . . . When the lamp begins to flicker, the green patch and the shadows come to life, and are set in motion, as though by the wind. It is stuffy. There is a smell of cabbage soup, and of the inside of a boot shop.

The baby is crying. For a long while he has been hoarse and exhausted with crying; but he still goes on screaming, and there is no knowing when he will stop. And Varka is sleepy. Her eyes are glued together, her head droops, her neck aches. She cannot move her eyelids or her lips, and she feels as though her face is dried and wooden, as though her head has become as small as the head of a pin.

"Hush-a-bye, my baby wee," she hums, "while I cook the groats for thee. . . ."

A cricket is churring in the stove. Through the door in the next room the master and the apprentice Afanasy are snoring. . . . The cradle creaks plaintively, Varka murmurs—and it all blends into that soothing music of the night to which it is so sweet to listen, when one is lying in bed. Now that music is merely irritating and oppressive, because it goads her to sleep, and she must not sleep; if Varka—God forbid!—should fall asleep, her master and mistress would beat her.

The lamp flickers. The patch of green and the shadows are set in motion, forcing themselves on Varka's fixed, half-open eyes, and in her half-slumbering brain are fashioned into misty visions. She sees dark clouds chasing one another over the sky, and screaming like the baby. But then the wind blows, the clouds are gone, and Varka sees a broad high road covered with liquid mud; along the high road stretch files of wagons, while people with wallets on their backs are trudging along and shadows flit backwards and forwards;

on both sides she can see forests through the cold harsh mist. All at once the people with their wallets and their shadows fall on the ground in the liquid mud. "What is that for?" Varka asks. "To sleep, to sleep!" they answer her. And they fall sound asleep, and sleep sweetly, while crows and magpies sit on the telegraph wires, scream like the baby, and try to wake them.

"Hush-a-bye, my baby wee, and I will sing a song to thee," murmurs Varka, and now she sees herself in a dark stuffy hut.

Her dead father, Yefim Stepanov, is tossing from side to side on the floor. She does not see him, but she hears him moaning and rolling on the floor from pain. "His guts have burst," as he says; the pain is so violent that he cannot utter a single word, and can only draw in his breath and clack his teeth like the rattling of a drum:

"Boo—boo—boo—boo. . . ."

Her mother, Pelagea, has run to the master's house to say that Yefim is dying. She has been gone a long time, and ought to be back. Varka lies awake on the stove, and hears her father's "boo—boo—boo." And then she hears someone has driven up to the hut. It is a young doctor from the town, who has been sent from the big house, where he is staying on a visit. The doctor comes into the hut; he cannot be seen in the darkness, but he can be heard coughing and rattling the door.

"Light a candle," he says.

"Boo—boo—boo," answers Yefim.

Pelagea rushes to the stove and begins looking for the broken pot with the matches. A minute passes in silence. The doctor, feeling in his pocket, lights a match.

"In a minute, sir, in a minute," says Pelagea. She rushes out of the hut, and soon afterwards comes back with a bit of candle.

Yefim's cheeks are rosy and his eyes are shining, and there is a peculiar keenness in his glance, as though he were seeing right through the hut and the doctor.

"Come, what is it? What are you thinking about?" says the doctor, bending down to him. "Aha! Have you had this long?"

"What? Dying, your honor, my hour has come. . . . I am not to stay among the living. . . ."

"Don't talk nonsense! We will cure you!"

"That's as you please, your honor, we humbly thank you, only we understand. . . . Since death has come, there it is."

The doctor spends a quarter of an hour over Yefim, then he gets up and says:

"I can do nothing. You must go into the hospital. There they will operate on you. Go at once. . . . You must go! It's rather late, they will all be asleep in the hospital, but that doesn't matter, I will give you a note. Do you hear?"

"Kind sir, but what can he go in?" says Pelagea. "We have no horse."

"Never mind. I'll ask your master, he'll let you have a horse."

The doctor goes away, the candle goes out, and again there is the sound of "boo—boo—boo." Half an hour later someone drives up to the hut. A cart has been sent to take Yefim to the hospital. He gets ready and goes. . . .

But now it is a clear bright morning. Pelagea is not at home; she has gone to the hospital to find what is being done to Yefim. Somewhere there is a baby crying, and Varka hears someone singing with her own voice:

"Hush-a-bye, my baby wee, I will sing a song to thee."

Pelagea comes back; she crosses herself and whispers: "They put him to rights in the night, but towards morning he gave up his soul to God. . . . The Kingdom of Heaven be his and peace everlasting. . . . They say he was taken too late. . . . He ought to have gone sooner. . . ."

Varka goes out into the road and cries there, but all at once someone hits her on the back of her head so hard that her forehead knocks against a birch tree. She raises her eyes and sees, facing her, her master, the shoemaker.

"What are you about, you scabby slut?" he says. "The child is crying, and you are asleep!"

He gives her a sharp slap behind the ear, and she shakes her head, rocks the cradle, and murmurs her song. The green patch and the shadows from the trousers and the baby clothes move up and down, nod to her, and soon take possession of her brain again. Again she sees the high road covered with liquid mud. The people with wallets on their backs and the shadows have lain down and are fast asleep. Looking at them, Varka has a passionate longing for sleep; she would lie down with enjoyment, but her mother Pelagea is walking beside her, hurrying her on. They are hastening together to the town to find jobs.

"Give alms, for Christ's sake!" her mother begs of the people they meet. "Show us the divine mercy, kindhearted gentlefolk!"

"Give the baby here!" a familiar voice answers. "Give the baby here!" the same voice repeats, this time harshly and angrily. "Are you asleep, you wretched girl?"

Varka jumps up, and, looking round, grasps what is the matter: there is no high road, no Pelagea, no people meeting them, there is only her mistress, who has come to feed the baby, and is standing in the middle of the room. While the stout, broad-shouldered woman nurses the child and soothes it, Varka stands looking at her and waiting till she has done. And outside the windows the air is already turning blue, the shadows and the green patch on the ceiling are visibly growing pale, it will soon be morning.

"Take him," says her mistress, buttoning up her chemise over her bosom; "he is crying. He must be bewitched."

Varka takes the baby, puts him in the cradle, and begins rocking it again. The green patch and the shadows gradually disappear, and now there is nothing to force itself on her eyes and cloud her brain. But she is as sleepy as before, fearfully sleepy! Varka lays her head on the edge of the cradle, and rocks her whole body to overcome her sleepiness, but yet her eyes are glued together, and her head is heavy.

"Varka, heat the stove!" she hears the master's voice through the door.

So it is time to get up and set to work. Varka leaves the cradle, and runs to the shed for firewood. She is glad. When one moves and runs about, one is not so sleepy as when one is sitting down. She brings the wood, heats the stove, and feels that her wooden face is getting supple again, and that her thoughts are growing clearer.

"Varka, set the samovar!" shouts her mistress.

Varka splits a piece of wood, but has scarcely time to light the splinters and put them in the samovar, when she hears a fresh order:

"Varka, clean the master's galoshes!"

She sits down on the floor, cleans the galoshes, and thinks how nice it would be to put her head into a big deep galosh, and have a little nap in it. . . . And all at once the galosh grows, swells, fills up the whole room. Varka drops the brush, but at once shakes her head, opens her eyes wide, and tries to look at things so that they may not grow big and move before her eyes.

"Varka, wash the steps outside; I am ashamed for the customers to see them!"

Varka washes the steps, sweeps and dusts the rooms, then heats another stove and runs to the shop. There is a great deal of work: she hasn't one minute free.

But nothing is so hard as standing in the same place at the kitchen table peeling potatoes. Her head droops over the table, the potatoes dance before her eyes, the knife tumbles out of her hand while her fat, angry mistress is moving about near her with her sleeves tucked up, talking so loud that it makes a ringing in Varka's ears. It is agonizing, too, to wait at dinner, to wash, to sew, there are minutes when she longs to flop onto the floor regardless of everything, and to sleep.

The day passes. Seeing the windows getting dark, Varka presses her temples that feel as though they were made of wood, and smiles, though she does not know why. The dusk of evening caresses her eyes that will hardly keep open, and promises her sound sleep soon. In the evening visitors come.

"Varka, set the samovar!" shouts her mistress.

The samovar is a little one, and before the visitors have drunk all the tea they want, she has to heat it five times. After tea Varka stands for a whole hour on the same spot, looking at the visitors, and waiting for orders.

"Varka, run and buy three bottles of beer!"

She starts off, and tries to run as quickly as she can, to drive away sleep.

"Varka, fetch some vodka! Varka, where's the corkscrew? Varka, clean a herring!"

But now, at last, the visitors have gone; the lights are put out, the master and mistress go to bed.

"Varka, rock the baby!" she hears the last order.

The cricket churrs in the stove; the green patch on the ceiling and the shadows from the trousers and the baby clothes force themselves on Varka's half-opened eyes again, wink at her and cloud her mind.

"Hush-a-bye, my baby wee," she murmurs, "and I will sing a song to thee."

And the baby screams, and is worn out with screaming. Again Varka sees the muddy high road, the people with wallets, her mother Pelagea, her father Yefim. She understands everything, she recognizes everyone, but through her half sleep she cannot understand the force which binds her, hand and foot, weighs upon her, and prevents her from living. She looks round, searches for that force that she may escape from it, but she cannot find it. At last, tired to death, she does her very utmost, strains her eyes, looks up at the flickering green patch, and listening to the screaming, finds the foe who will not let her live.

That foe is the baby.

She laughs. It seems strange to her that she has failed to grasp such a simple thing before. The green patch, the shadows, and the cricket seem to laugh and wonder too.

The hallucination takes possession of Varka. She gets up from her stool, and with a broad smile on her face and wide unblinking eyes, she walks up and down the room. She feels pleased and tickled at the thought that she will be rid directly of the baby that binds her hand and foot. . . . Kill the baby and then sleep, sleep, sleep. . . .

Laughing and winking and shaking her fingers at the green patch, Varka steals up to the cradle and bends over the baby. When she has strangled him, she quickly lies down on the floor, laughs with delight that she can sleep, and in a minute is sleeping as soundly as the dead.

There are so many people who contributed to the making of this book:

Elyse Cheney

First and foremost, I'd like to thank all of the contributors for their patience, their excellent contributions, their support, and their enthusiasm. Without them, this book would still be just an idea tossed around at a lunch meeting. I'd also like to thank Connie Clausen, whose expert advice and constant support gave me courage and guidance. I very much appreciate what you've done for me.

Thank you Rob Robertson, who inherited the project once Wendy left. You took this on as if it were your own. Your good-hearted proddings had just the right light touch. I couldn't have asked for a better editor.

I'd also like to thank Anthony Schneider for hanging by me when the going got tough and for his love and laughter.

To my father, Nat Cheney, my mother, Karen Cheney, and my sister, Meredith Cheney: Thank you for always being there.

Finally, I'd like to thank all those sleepless nights that provided the inspiration for these stories.

Wendy Hubbert

Huge thanks to our contributors, who responded with such enthusiasm and creativity—not to mention alacrity—to the subject of sleeplessness.

Thanks go as well to Doubleday and Rob Robertson, Lori Lipsky, and Scott Moyers for their invaluable initial and continued support.

A C K N O W L E D G M E N T S

"Music at Night" © 1996 by Tom Beller.

"Dreams" © 1996 by Tim Cahill.

"Mr. Morning" © 1989 by Siri Hustvedt. Originally published in *The Ontario Review*, issue 30, 1989.

"Waking Up" © 1996 by E. Annie Proulx.

"Sleeping and Waking" © 1931 by F. Scott Fitzgerald. Originally published in *The Crack-Up*, Charles Scribner's Sons.

Elyse Cheney is a literary agent in New York City.

Wendy Hubbert is a book editor in New York City.